Wildlife & Woodlot Management

Wildlife & Woodlot Management

A Comprehensive Handbook for Food Plot and Habitat Development

Monte Burch

Skyhorse Publishing

Dedication

To two close friends from the Missouri Department of Conservation. David Pitts got me interested and started in wildlife management back in the '70s when he was a Landowner Specialist with the MDC. The late Carl Conway also worked with landowners in wildlife management and was instrumental in helping me with various programs. Carl was also a very good friend and college fraternity brother.

Skyhorse Publishing books may be purchased in bulk at special discounts for sales promotion, corporate gifts, fund-raising, or educational purposes. Special editions can also be created to specifications. For details, contact the Special Sales Department, Skyhorse Publishing, 307 West 36th Street, 11th Floor, New York, NY 10018 or info@skyhorsepublishing.com.

Visit our website at www.skyhorsepublishing.com.

10 9 8 7 6 5 4 3 2 1

Library of Congress Cataloging-in-Publication Data is available on file.

ISBN: 978-1-62087-786-9

Printed in the United States of America

Table of Contents

Introduction

More and more people are interested in attracting wildlife to their property—from those with tiny backyard patios to those who own vast tracts of land. Many people manage their land for hunting and fishing; others manage simply to attract wildlife to watch and enjoy. Wildlife management can be simple and economical or take lots of hard work and money, depending on the owner's goals, the land type, the existing habitat, and the species the landowner wishes to attract.

I grew up on a farm with lots of wildlife, although back in the '60s there were no white-tailed deer or turkeys in our area. Thanks to farsighted wildlife management plans, those two very popular species have been reintroduced to much of the United States and have flourished.

In the early '70s my wife, Joan, and I purchased a run-down hill farm in the Ozarks. Abandoned for several years, the land was worn out. There was little grass, the timberlands had been grazed to bare earth, and there were few game birds and animals. We learned that the Missouri Department of Conservation offers help to landowners wanting to create wildlife habitat, and we met David Pitts. At the time, Dave was a field service agent for the Department, and his job was to help landowners manage wildlife habitat. Over the years, Dave's sage advice and friendship have been invaluable. We also received advice from local county extension agents, Soil Conservation Service personnel, and many others. At that time very few private individuals were managing habitat specifically for wildlife, and our neighbors thought we were a little crazy to put so much time and effort into something with no monetary reward.

We began planting food plots and improving and extensively managing pastures, haylands, and croplands. We also made widespread timber stand improvements, planted trees and shrubs to feed a variety of wildlife, seeded fields with warm-season grasses, fenced creeks, and constructed ponds. Many of these practices made our farm not only more profitable, but also much more attractive to wildlife. Over the years I have made numerous notes and collected information and ideas from anyone I could talk to about wildlife habitat management. When companies such as the Whitetail Institute began producing wildlife-specific seeds and plants, I volunteered as a test planter. Since then I've tested numerous seeds and continue to test them. When ATVs became popular habitat-management tools, I experimented with the accessories available for them. They can be used for such diverse tasks as planting food plots, mowing, pond dams, and setting deer stands.

These days our farm is a wildlife haven. From our office windows we see wild animals every day. Often work is forgotten for a few moments when we spot a really big deer, a gobbler strutting, wood ducks landing on a pond, or an unusual bird at one of our feeders. We also have an abundance of just about every wild critter native to our area including rabbits, squirrels, quail, doves, and lots of non-game wildlife. Wood ducks and Canada geese nest on our property, and other migrating waterfowl use our seven ponds. These ponds provide lots of fishing, frogging, and family fun time. I'll never forget the amazement of our grandchildren one spring evening when we took a Kawasaki Mule drive through our property and they heard a pond full of spring peepers for the first time. How could such little things make such a big noise!

Wildlife habitat management is fun and can add to the enjoyment of your property, not to mention creating great hunting and fishing opportunities. I hope you and your family enjoy wildlife habitat management on your own property as much as we have.

THE LAND

CHAPTER 1

CHOOSING LAND

Before you can manage habitat for wildlife, you need to lease or own land. If you lease, be sure to get a long-term lease, because many wildlife habitat practices take a long time to show results. The dream of most people who enjoy hunting, fishing, or simply watching wildlife is to own property, but the old adage "They aren't making any more land" is even more applicable these days than when it was first said. With more land being gobbled up by urban sprawl, highways, and even big agriculture, less is available for wildlife. And not only is land for wildlife becoming more scarce, it's also becoming more costly. For instance, not too many years ago "rough"—hilly and unimproved—Ozark land went for $500 an acre. At the time of this writing, that same land goes for well over $1,000 an acre and the price is climbing! It seems that everyone wants a wildlife haven or their own acre in the country.

Proper habitat management is the key to hunters, anglers, and landowners desiring to attract and keep wildlife for hunting, fishing, or just watching.

Planning

If you dream of owning your own wildlife haven, first you must set goals and make a plan. How much can you afford? Will you live on the land? How much does land cost in your desired location? How much will insurance and taxes cost?

Then decide what type of wildlife you're interested in. In some cases you may be limited by existing habitat or topography, but often the habitat and management for one species may also be suitable for others. For example, managing a forest with openings for deer also benefits turkeys, quail, and numerous songbirds as well as small game such as squirrels and rabbits. And a pond or lake created and managed for fishing can also attract waterfowl and other wildlife.

We were very lucky when we purchased our farm back in the '70s. Located "where the Ozarks meet the prairie," as a nearby small town describes the area, the land is extremely diversified, with hardwood timber, open prairie, a creek, and a small marsh.

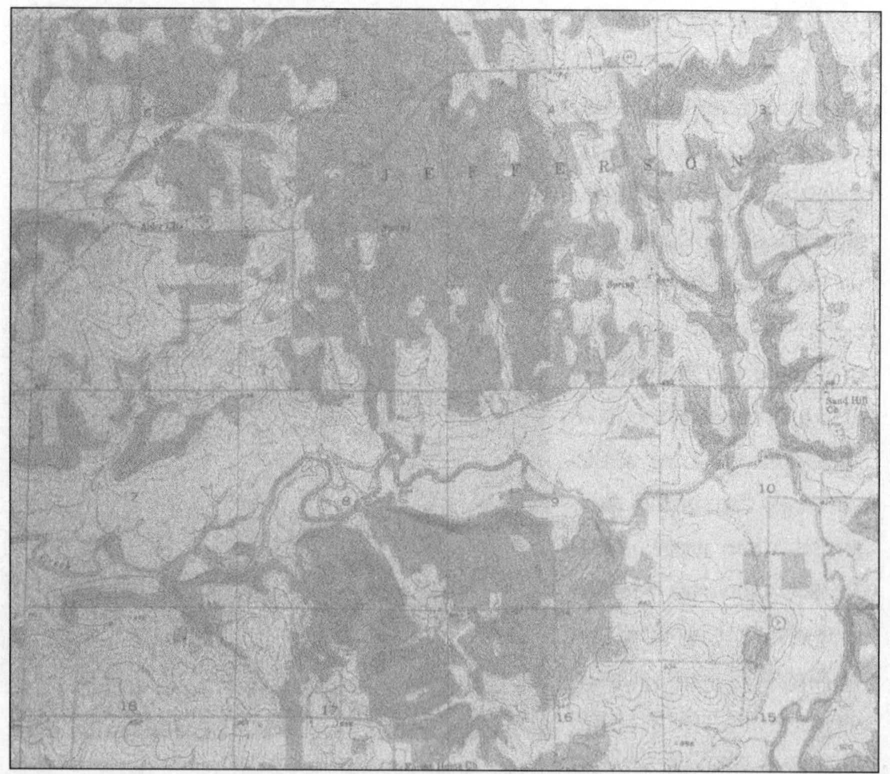

Tabletop planning is the first step whether in purchasing land or planning practices on existing land. A topographical map details the terrain.

When looking at a property to buy, two tools are valuable in determining wildlife potential—a topographical map and an aerial photo. These can illustrate the topography—whether the land is flat, sloping, hilly, or mountainous—and whether it has streams, marshes, or other features. The aerial photo will also show vegetation types. In addition, a county plat map can tell you who owns land you're interested in. Other factors to determine are the average yearly rainfall, availability of water, soil types, and weather conditions, because some management practices can be successful only with proper rainfall and soil types. If you dream of having your own fishing lake, for example, the soil types and topography of the area will determine the suitability of lake or pond construction.

Can your land can be monetarily productive as well as managed for wildlife? Usually a well-managed farm, ranch, pine plantation, or timberland can be both a wildlife haven and a money-maker. In fact, some studies show that well-managed timberlands can, over the long run, be more profitable than investing in the stock market. By managing timberlands you can make money selling such diverse products as chipwood, firewood, saw logs, and Christmas trees. Even unusual products such as grapevine wreaths, rustic furniture, mushrooms, ginseng, and other plants can be money-makers.

On prairie and agricultural lands, agricultural crops, hay, and livestock can all be managed along with wildlife. Even if you're not interested in doing these things yourself, leasing the land to a farmer or neighbor can benefit both parties. Or you may wish to sharecrop, allowing a neighbor to cut hay or plant and harvest corn or other crops. If you lease or sharecrop, make sure each party understands the agreement, and put it in writing. Also be sure that the other party knows that your primary interest is managing for wildlife, because some farming practices are not conducive to attracting wildlife.

CHAPTER 2

ASSESSING LAND

Regardless of whether you've just purchased land or already own land you wish to improve for wildlife, it's important to assess the overall habitat management potential of the property. Follow this assessment with a management plan. Your aerial photo and topographical map can be used in planning sessions, but you'll also need to get out and walk the land. Don't hesitate to ask for help. Plenty of experts will give valuable free advice, and a number of federal and state wildlife habitat improvement programs offer funding. Begin by contacting your state fish and wildlife agency. You may also wish to contact the Natural Resources Conservation Service (NRCS), county Soil Conservation Service (SCS), county University Extension Service, and state and private foresters.

IDENTIFYING HABITAT TYPES

Begin your assessment by identifying and marking six broad habitat types on the aerial photo:

1. **Upland woodlands:** Forests, or areas overgrown with trees with a canopy greater than ten percent.
2. **Bottomland hardwoods:** Forested bottomlands or wood swamps. This may also include tree-lined oxbows.
3. **Non-forested wetlands:** Potholes; marshes; sloughs; low, wet grassy areas; and shallow, water-logged depressions.
4. **Pasture and haylands:** Includes native prairies.
5. **Old fields:** Agricultural fields or pastures abandoned for more than two years and having less than ten percent canopy of overstory trees.
6. **Croplands:** Fields planted to row crops and small grains.

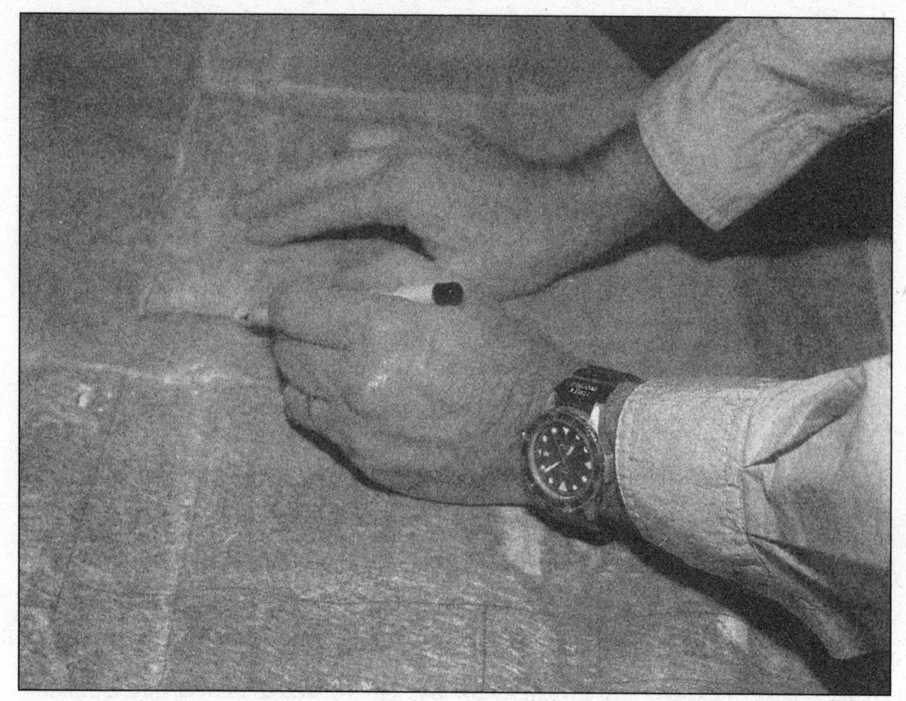

An aerial photo is a very important tool. It shows vegetation as well as other important features.

Once you've marked these areas on the aerial photograph, estimate the acreage of each habitat type. Now it's time for fieldwork. With a notebook, walk the property, examine the different areas, and make notes on the suitability of existing habitat for desired species as well as for possible improvement practices. Talking to an expert in the field can be a great help in this step. And before you make these evaluations, read the rest of this book to learn about management practices for different habitat types and the wildlife species you wish to attract.

UPLANDS

Concealment cover. Estimate the amount of dense or shrubby areas, brush piles, rock piles, fallen logs, and other cover. Include dense, shrubby draws extending into at least fifty percent of a field.

Edge. Examine the edges of fields or borders of the field. These include hedgerows, overgrown fencerows, and strips of vegetation between habitat types. Note whether the edges are straight or irregular. Determine the number of nest or roost trees (these include either dead or live trees of greater than six inches diameter at breast height [DBH] with cavities), including coniferous trees such as pine or red cedar. These are all used by doves.

Vegetative cover. Vegetative cover less than twenty percent will not supply enough cover and food for many species. Canopy coverage of shrubs and herbaceous vegetation that is six inches to four feet tall is preferred by white-tailed deer. Cover six to eighteen inches tall is preferred by other species, including quail, rabbits, and turkeys. If the area has more than sixty percent coverage of shrubs and herbaceous vegetation, however, it may be too thick for ground-nesting birds and small mammals to walk through. Note the types of cover—cool-season grasses, warm-season grasses, legumes, or a mixture of plants.

Past grazing or haying pressure. Determine past grazing or haying pressure on the area as well as past flooding or burning. Moderate pressure leaves three to six-inches of cool-season grasses and eight to twelve inches of warm-season grasses over winter. Name and estimate the percentage of legumes, including clovers, in the grasslands. Finally, note the grassland species, including existing forbs.

Croplands

Determine the normal or past cropping rotations, including rotations into grass. Also determine land fertility.

Woodlands

Oak trees. Estimate the percentage of black and white oak groups in the forest, since they are the most important mast-producing trees.

Size class. Determine the size class of the woodlands, defined by the diameter at breast height (DBH). The size classes are:

1. *Old growth:* determine the percentage of trees greater than sixteen inches DBH.
2. *Saw timber:* trees greater than nine inches DBH.
3. *Pole timber:* two- to nine-inch DBH.
4. *Regrowth:* zero to two-inch DBH.

Canopy. An "open" canopy has less than fifty percent coverage. A "closed" canopy has greater than fifty percent coverage.

Nest and roost trees. Determine the percentage or number of nest and roost trees. These include either dead or live trees of greater than six inches DBH with cavities.

Understory. Determine the density and makeup of the forest understory. If there are more than four stems per square yard, walking through the forest will be difficult. These areas are, however, great for ruffed grouse.

Wetlands

Water availability. Determine the fall and winter water availability.

Flooding. Estimate the amount of land that can be flooded from one to eighteen inches deep either naturally by rainfall or manually by artificial means. If it is more than eighteen inches, puddle ducks can't tip up to feed.

Wetland plants. Determine the types and estimate the percentages of existing wetlands plants.

Winter cover. Estimate the percentage of winter cover, including woody vegetation and/or emergent plants that can provide protection from the weather.

Habitat Percentages

After you've assessed the habitat types on your property, it's useful to also determine the percentage of habitat types within a two-mile radius of your land. This allows you to manage for wildlife in conjunction with habitat surrounding you; perhaps you can offer something that is not already available.

Uplands. Estimate the percentage of native grasses within a two-mile radius and also the percentage of pasture and haylands within the same area.

Croplands. Determine former cropping practices and existing crops within a two-mile radius.

Determining the types of habitat that surround your property will help you plan what you will plant, develop, or enhance on your land. COURTESY FIDUCCIA ENTERPRISES

Woodlands. Estimate the amount of forest cover within a two-mile radius of the tract being examined. (Deer and turkey numbers are directly related to the amount of forest cover.)

Wetlands. Estimate the percentage of forested bottomlands and unforested wetlands within a two-mile radius of your field or wetland. (Ducks, such as mallards, are more attracted to large wetlands or groups of wetlands than to isolated ones.) Also determine the distance to the nearest large reservoir or waterfowl refuge.

Overall percentages. Determine the percentage of croplands, native warm-season grasses, woodlands, and wetlands within a two-mile radius of your property. The amount and types of upland habitat, including grasses and croplands, determines whether quail, rabbits, pheasants, prairie chickens, and doves will be attracted to the land. The amount and types of woodlands determines the suitability for white-tailed deer, turkeys, squirrels, and ruffed grouse. The amount and types of wetlands determines the suitability for waterfowl. As you can guess, diversity provides for more wildlife species.

With this assessment you will be able to determine what wildlife habitat management practices are suitable for your property. You may discover that some practices are not practical on your property. You may find that your land is capable of attracting or holding certain wildlife species, but the surrounding land is not. But you can, with the right property, money, and hard work, manage the habitat for a wide variety of species or manage for your preferred species.

CHAPTER 3

PLANNING FOOD PLOTS

Creating food plots (sometimes called "green fields" in the South) can be a very important facet of a landowner's wildlife management plan. Wildlife have four basic needs for survival: food, water, shelter from the elements and from predators, and space. These needs become increasingly important in winter, when animals need more energy to survive and food is often harder to find. Food plots can play a crucial role in winter survival, offering both food and shelter to a variety of wild animals. If you do nothing else in the way of management, providing food plots can attract wildlife to your land as well as provide food when it's most needed.

Although growing food plots is a relatively new idea to many hunters and landowners, I've been experimenting with them for about thirty years on our Missouri farm. Since I have a full, working farm, I've brought over many

Food plots can be extremely important wildlife habitat for any number of species. COURTESY FIDUCCIA ENTERPRISES

practices from my cow-calf and cash-crop alfalfa operation. I began with advice from Dave Pitts, then with the Missouri Department of Conservation, and planted green browse plots of ladino clover for deer and rabbits and annual plots of milo and soybeans for deer, turkeys, and upland birds. Each plot brought more wildlife. These days I'm constantly experimenting with new seeds, redoing plots and creating new ones. For example, I cleared four acres of nuisance trees—Osage orange and hardwood brush—and planted the area in a combination of warm-season grasses, the Whitetail Institute's Imperial Whitetail Clover, and Alfa Rack. Joan filled her antlerless tag the first day of the season from the spring-planted food plot and then took a ten-point buck—her first buck—from the same food plot the second day of the season.

Food plots can provide nourishment for a variety of wildlife. The wildlife attracted depends on the type of food in the plot. In recent years many landowners and hunters with leased land have planted food plots to attract and hold deer and, just as importantly, provide the supplemental nutrition to help the deer reach their optimum body and rack size. These food plots will attract wildlife other than deer as well.

"One reason for the growing food plot popularity is that more people are understanding nutrition as a manageable aspect," said biologist and deer expert Dr. Grant Woods from South Carolina. "We can manage three factors in a wild, free-ranging herd—age, sex ratio, and nutrition. The first two are totally dependent on the trigger finger. Nutrition is forest management—clear-cuts or thinning; prescribed fire or herbicide treatment; and agricultural practices including food plots. Food plots can be very important in many parts of the country. The average forage intake needed per deer per year is about two thousand pounds of dry matter. There's a wide variance, of course, with a fawn needing, say, eight hundred pounds, while a mature buck may require four thousand pounds, but the average is two thousand pounds per deer. Food plots can provide five thousand to ten thousand pounds of forage per acre per year, depending on the plants and the rainfall."

Steve Scott, with Whitetail Institute, the company that introduced Imperial Whitetail Clover, the first commercial deer food-plot product, in 1988, says his company has seen interest in food plots increase tenfold in the past few years. "We're finding that many people are having as much fun managing their land for deer and other wildlife as they are hunting."

You can't simply throw seed out on the ground and expect food plots to succeed. Successful plots require both time and money. Follow these eight steps and you'll be on your way to success: planning, land clearing or preparation, soil testing, liming, fertilizing, preparing the seed bed, seeding, and maintaining.

PLANNING

LOCATION

Determine your food plot needs. What wildlife do you wish to attract and/or grow? Examine your property on an aerial photo and a topographical map to determine possible food plot locations. The aerial photo shows existing vegetation, while the topographical map illustrates where level areas exist that may be suitable for planting. Determine whether overgrown fields or areas of brush or low-grade timber could be cleared to create food plots. If the land is agricultural, identify areas that can be set aside to provide foods other than agricultural crops at different times of the year from when the agricultural crops are grown. Food plots can also be planted along logging and interior farm roads, but don't place them along public roads, because they may attract poachers and can also present a hazard to motorists. The best food plot locations are fairly accessible for tillage equipment, open, and tillable. Although open, they should also be close to good cover and escape routes. Good locations include along

Food plots must be located near cover, but away from public roads and property boundaries. The size and shape can vary, but long and narrow plots of at least one-half acre are best.

streams, timber edges, or timber clearings; near brushy draws; and in the corners of shrubby fencerows. If you plant them near timber edges, locate them at least twenty yards from the timber edge to reduce competition from trees and prevent overshading. The area between the food plot and the timber edge can be planted in shrubs to provide escape and cover for animals coming into the field.

If you have tracts of woodlands or timber, you can also create food plots in them. Woodland food plots offer the ideal situation for deer and turkeys: the woodlands have food and shelter and the food plots create diversity, more edge, and food. Both deer and turkeys love edge habitats, which are often called "wildlife openings" by biologists and wildlife managers. Old clearings, log loading yards, and other openings in timber may offer food plot possibilities. Abandoned homesteads are also great food plot locations. Right-of-ways can also be used, but check with lessors about legal use of such areas.

In pine plantations, one method of creating a food plot is to make a "hub-and-spoke" clearing design. To do this, make a center opening of half an acre or larger and then run spokes from this center hub. The spokes are usually about thirty yards wide and up to one hundred yards long. Many people build a shooting house or tower in the center of the hub.

If you clear timberland, choose areas that have timber of poor value. Areas with firewood potential are good choices in some parts of the country. If you log timberland for sale or thinning, a spot used as a landing for loading logs can later be turned into a food plot.

If hunting is the major reason for the food plot, it's important to plant it in a huntable location. Locate it not only for hunting success but also for safety. For instance, a food plot with the only stand location directed toward a public road or houses presents a safety problem, and plots that are unapproachable due to prevailing wind direction are difficult to hunt.

All of the food plots on our property are back some distance from neighboring fences and property lines. Food plots on or near property lines may tempt others to shoot onto your property, and if you shoot an animal and it then runs back onto the neighbor's property, you may have a problem.

Soil Type

Food plot soil should be tillable but not highly erodible. Bottomland, the flatter tops of ridges and hillsides, and along the contour of gentle slopes are all possible locations. Since good soil grows the most productive food plots, bottomland, along-side streams, and other areas with fairly deep soil are the best choices. You can plant on thin hilltop soils, but these areas tend to have problems in times of drought. Plant the plots on relatively flat terrain, not sloping hillsides, as tilling hillsides can cause serious erosion. You can add diversity in hillside clearings, however, by planting permanent vegetation such as warm-season natural grasses that will hold the soil in place.

Deep, fairly heavy soil that holds moisture is best, but even marginal soils—those that are thin, rocky, semi-arid, or somewhat sandy—can grow food plots if you choose the correct seed and practice proper soil management.

Size, Number, and Shape of Plots

Size. The size of each food plot depends on the characteristics of your property, rainfall, deer density, and how large the suitable areas are. Food plots should be at least half an acre, and you may need to plant plots of one, two, three, or even more acres in areas with high deer densities. Deer can quickly overbrowse small plots, which lowers the attractiveness of the plot, causes problems with plant rejuvenation, and shortens the plot's lifetime. It is, however, better to have numerous small plots than one huge plot.

Number. The number of food plots depends on all of the above factors as well as the availability of other nearby food sources, both natural and agricultural. Many experts suggest a minimum of one food plot per forty acres in heavily timbered locations. I have a dozen or so plots on three hundred acres. Wildlife management consultant Larry W. Varner, Ph.D., from Texas, says the number of plots should be determined by the number of deer, the amount of rainfall, and the production potential of the species planted. "When deer density is one deer to twenty acres or more, I recommend that one to two percent of the total area be in food plots. With higher deer densities—one deer to ten acres or less—I recommend three to five percent. Another way to figure acreage is to go by average annual rainfall. In areas with thirty inches or more per year, I recommend about 0.15 acres per deer. In areas with twenty-five inches or less per year, you'll need 0.30 acres per deer."

Shape and orientation. The ideal shape of a plot is long and narrow, following the edge of the cover, rather than square and blocky. This allows a greater edge-to-interior ratio, providing more opportunities for wildlife to visit the plot yet have a quick escape into cover. Big square plots usually have visitation only around the edges except by turkeys and upland birds. The width of the opening into the food plot must be at least one and one-half times the height of any adjoining trees. This allows sunlight to reach most of the plot and encourages better growth. If possible, openings should also be oriented to face in an east/west direction to allow for maximum sunlight.

Field Work

Once you've determined the size, number, and location of your plots, visit the sites and evaluate their feasibility by examining the soil and determining how much clearing, herbicide application, and/or tillage will be needed. You should also plot roads or routes to reach the site for both planting and hunting. Take soil samples from each of the sites and have them analyzed at your local county extension office. Let them know what types of seed you'll be using, because different seeds have different soil requirements.

Costs

If you're planning your first food plots, you should know that successful food plots are not cheap. Once a plot has been established, including land clearing, initial liming, fertilizing, seeding, and so forth, the average annual cost to maintain it can run from $90 to $150 per acre. In areas of high deer density, you may also have to install an electric fence around the plots until they're established, at an additional cost.

Types of Food Plots

Food plots can be planted to provide a variety of winter, summer, spring, and fall foods or year-round foods, and several types of plants may be used. These include annuals such as wheat, oats, corn, milo, soybeans, and various pea species; and annual brassicas and perennial legumes such as clovers and alfalfa. Plots can be planted in the spring or fall, but fall planting is dependent on the species selected and the availability of rain. Spring and fall plots offer wildlife different foods at different times of the year.

Larry Varner offers advice on how much to plant in spring and fall: "Another consideration is how much of the total food plot plan should be in summer vs. winter food plots. For each fifteen acres of food plot area, I would plant ten acres (sixty-six percent) in summer plots and five acres (33 percent) in winter plots. The reasons for this are twofold: (1) winter food plot species (wheat, oats, etc.) tend to be more productive than most spring-planted species and (2) improved nutrition in the spring and summer has the greatest impact on subsequent antler production and fawn growth."

Food plots may be planted with annuals or perennials. Clover is one of the most popular perennial plants and it attracts all types of wildlife.

Timothy E. Fulbright, Ph.D., with the Caesar Kleberg Wildlife Research Institute, Texas A & M University–Kingsville, says that cool-season food plots of oats, wheat, and triticale, along with legumes such as hairy vetch, Austrian winter peas, alfalfa, and clover, should provide nutritious forage from November through April. Warm-season crops such as lablab, cowpeas, soybeans, and milo should provide forage from early spring through late fall.

Annuals, of course, must be replanted each year, while perennials last for a number of years. In most instances a combination of perennials and annuals provides the best food plots. I use both annuals and perennials in my plots primarily because I test a lot of new products, but also because the different varieties offer wildlife a smorgasbord of food throughout the year.

Clover is one of the most popular food-plot seeds, especially for deer and turkeys. It has high palatability and provides good nutrition, and once established, the plot may stay for several years. One Imperial Whitetail Clover food plot on our property lasted for almost ten years. "Clover provides the best of both worlds," said Whitetail Institute's Steve Scott. "You are not going to find anything that is more preferred year-round. We'd be the first to admit that when white oak acorns are falling, it doesn't really matter what you've got out, the deer are going to the acorns. But white oak acorns are available for a very short time. And clover provides the highest possible nutritional value at the right times of the year—the spring and summer months when you have antler development and third-trimester pregnant does."

"There is no single 'magic bean,' including BioLogic," explained Grant Woods, one of the developers of the Mossy Oak products. "A good food-plot system matches the needs of the deer as their annual cycles change. In the spring deer need high protein and a lot of tonnage. As the summer develops into July and August, deer need moderate protein and a lot of energy. By late fall and carrying through the winter, they don't need quite so much protein, but they need a high amount of energy. Specific plants provide those nutrients at certain times.

"Small grains, such as wheat and oats, are medium for quality and their digestibility is not that good, but they are good for energy. Plant these small grains in the fall. The brassicas are the most drought-resistant group of plants, and they produce the most tonnage per unit of effort with their huge leaf surface area and very small stem surface. The ideal food plot would have brassicas in the fall for their high energy and nutrients. Mix in a legume such as alfalfa or clover. Clovers don't do as much in the fall, but they really shine in the spring and summer. Clovers are the first to green up, and turkeys and deer both love them. Clovers carry through that early season when there is nothing to eat in the woods—the acorns are all gone and the animals have browsed all the edible twigs.

"If you have a low deer density, some of the more progressive legumes, like soybeans or peas, are great forage. They have little browse tolerance, however, and a high deer density can wipe out a small plot before the plants can grow any tonnage. The bottom line is that you need a blend. Another advantage of blends is that if you have four, five, or six species or cultivars, and something fails because it is too wet or too dry, one weak cultivar is usually balanced out by something that is strong in that area. When you plant a monoculture, you are really exposing yourself to either the vagaries of the weather or pests."

Seeds

Any number of plants can be used for food plots for fall hunting, but some plants are at their best early in the hunting season or until the first frost. Other plants last into the winter months and provide an attractant during the late muzzleloading and bow seasons.

Annual seeds include spring-planted varieties such as peas, some clovers, rapeseed, brassicas, corn, beets, soybeans, and the sorghums. Most of these plants offer food from early summer until a hard freeze, although some are hardier and will last through a few freezes and several are hardy in southern zones. Some of the brassicas I've tested have done pretty well in Missouri even in hard winters. Other annuals include winter wheat, triticale, and oats. These are planted in late summer or early fall and some can provide green foods throughout the winter. Oats don't last all winter, but winter wheat provides food throughout the winter and deer are attracted to it when other plants have died back.

The perennials include alfalfa, most clovers, chicory, ryegrass, lespedeza, and deer tongue grass. Although there is no single best "all-around" plant, clover comes the closest. A wide variety of red, crimson, and white clovers is available. Some varieties are the first plants to green up in the spring, providing food throughout the summer and fall and into winter until a hard freeze. White clover and blends of ladino clover are some of the most popular perennials. The legumes—such as most clovers and lespedeza—do not have to be replanted each year, and with proper maintenance, clover food plots can last several years. Legumes also attract turkeys that eat the many insects preying on these succulent plants.

You'll find many food-plot seeds available locally. Check with your county extension office or local seed dealers about the seeds that do best in your area. Some companies now sell seeds specifically designed for food plots. One of the first to market seeds for deer food plots was Whitetail Institute with its Imperial Whitetail Clover, a blend of seeds that includes early-starting varieties, varieties that germinate later, and varieties that are drought-resistant. Imperial Whitetail Clover is also preinoculated, eliminating the step of inoculating the seed to provide a better chance of success, a timesaver for first-time food plotters. I've been experimenting with Imperial Whitetail Clover for well over ten years with good luck except during extreme droughts.

Generic Food Plot Seeds

A wide range of generic agricultural seeds that can be used for food plots is available from local seed supply stores. The seeds go by a variety of names, so I won't attempt to enumerate them here. It's best to check with your county extension agent or local seed distributors about what works best in your area.

Corn. More than one hundred species of wildlife are known to eat corn. Deer, turkeys, and upland game love it. And small critters, such as raccoons and squirrels, can quickly reduce a corn patch to a barren field of green stalks. Corn is one of the best sources of carbohydrates and fat, but it's low in protein. For this reason corn provides good nutrition in fall and winter when other sources may be unavailable. An annual, corn is planted in the spring. Use field corn, not sweet or silage varieties. *Disadvantages:* Food plots for corn must be fairly large to prevent overeating. Corn is also somewhat more drought-prone than other food plot plants, and lots of fertilizer is needed for its production. It can be broadcast, but does best when row-planted.

Grain sorghum. Often called "milo," this plant is a close relative of corn. The leaves and stalks are relished by deer early in the season, and birds love the seeds the plant produces. Milo is also an excellent overwintering plant for many species of wildlife. It is an annual and is planted in the spring. *Disadvantages:* It has high fertility requirements, and it is usually drilled or row-planted.

Clovers. The clovers, including white, ladino, red, and crimson, are excellent food-plot seeds. The white, ladino, and red clovers are perennial. Red clover is fairly short-lived; it usually lasts two to three years. Some varieties of white or ladino clover can live for more than five years under the right conditions. Crimson clover is a fall-seeded annual. All the clovers are legumes, which means that they fix nitrogen in the soil. They are also primarily cool-season crops, which means that they produce forage in early spring and then again in the fall. Clover is fairly easy to establish. Broadcast-seeding on prepared food plots works best, although no-till planting on herbicide-killed or burned areas also works. Deer love the clovers, and turkeys, quail, and other birds not only eat the plants but relish the numerous insects they attract. Clovers can be planted in spring or fall. *Disadvantages:* The clovers tend to have shallow root systems and are affected by drought, some varieties are more drought-resistant than others.

Wheat. Wheat will grow almost anywhere in the United States and is one of the most important food-plot plants. A cool-season grass that grows up to 4 feet tall, it provides nutritious, tender forage in the fall, through the winter, and into spring. In late spring and summer the mature seeds in the seedheads provide high levels of protein. Wheat is an annual and is broadcast or drilled in early fall. Some experts like to combine wheat with legumes such as clovers or winter peas. *Disadvantages:* None.

Soybeans. A common agricultural legume, soybeans, especially the succulent young plants of summer, are loved by deer. Like all legumes, soybeans fix nitrogen in the soil and are often planted as a rotation crop with corn, which has high nitrogen requirements. Like corn, soybeans are high in carbohydrates. The soybean seeds are also readily eaten by turkeys, quail, and other upland game, especially during the winter months. Soybeans are annual and are usually drilled but can easily be broadcast-seeded. They are planted in late spring or early summer. *Disadvantages:* Soybeans have fairly high fertilizer requirements, and unless you plant a large plot, the deer will eat it to the ground before the plants have a chance to mature.

Alfalfa. Alfalfa is one of the most palatable, nutritious, and high-protein food-plot plants. Creeping or grazing alfalfa, which steadily renews itself and can last up to five years (longer with proper maintenance), has become increasingly popular for food plots. It has deeper roots than many other forage crops and can withstand drought better than the clovers. It can be planted in either spring or summer and is most commonly broadcast. It's fairly easy to seed, but it does require a well-tilled and prepared seed bed. *Disadvantages:* Plants are highly insect-prone (but this does attracts ground-feeding birds).

Other plants. Rapeseed, chicory, sugar beets, and plain old turnips are often used in hunt-attractant food plots for deer.

Commercial Food Plot Seeds Designed Primarily to Attract Deer

A number of commercially produced food plot products, designed primarily for deer but also used by other wildlife, are available. Many must be planted in the spring, although some are planted in the fall and some can be planted during either season depending on locale, temperature, and annual rainfall.

BioLogic. The BioLogic family of food-plot products was the result of the dedication of two individuals—Toxey Haas and Grant Woods—to managing white-tailed deer. Haas has a passion for chasing and raising whitetails, and Woods researches and manages deer herds.

- *New Zealand Premium Perennial Blend* provides a nutritious, highly palatable blend of forages throughout the growing season. The blend consists of perennial plants and annual brassicas. Plant in early fall in the South and in the spring in the North. As the brassicas die out in winter, broadcast BioLogic's Maximum, composed

Although local seed dealers carry most food plot seeds, a wide range of "specialty" food plot seeds are also available.

of one hundred percent New Zealand brassicas, directly onto the existing perennial crop during the early fall portion of the second growing season to maintain an optimum forage rotation.

- *Clover Plus* is a blend of New Zealand red and white clovers along with varieties of chicory. This blend produces high-quality forage, especially during the hot months of summer when other crops may be stressed. Plant in the early fall in the South and frost-seed or plant in the spring in the North.
- *Maximum* contains one hundred percent New Zealand brassicas that can be planted in either spring or fall. These brassicas yield as many as ten tons of forage per acre, with over thirty-eight percent crude protein. Planted in early fall, they will germinate in drier conditions than other fall food-plot crops. If planted in the fall, they must be planted thirty days before the first frost date; they can also be planted in the spring.
- *New Zealand Full Draw* is composed of a mixture of cultivars that offer nutrition from germination until the plot matures well after the hunting season. It is planted in early fall in the South or thirty to fifty days before the first frost date in the North.
- *Green Patch Plus* contains a blend of wheat, oats, clovers, and brassicas for large food plots. Plant it in the fall; it needs at least thirty days of growth before frost.

- *Biomass* produces high-quality protein through the spring and summer months, which maximizes antler growth and fawn development. Biomass requires daytime temperatures in the seventies or higher to ensure rapid germination, but it can be planted anytime during the spring after the last frost date and through the summer or in early fall as long as there is adequate soil moisture. When used as an early bow season attractant, plant it three weeks before bow season starts.

Whitetail Institute. Whitetail Institute was the first to design and distribute seed specifically for white-tailed deer, and they continue to produce a line of high-quality products for deer food plots, many of which are relished by other wildlife as well.

- *Imperial Whitetail Clover* does best in soils that hold moisture. It produces up to thirty-five percent protein year-round and lasts up to five years after a single planting. It's coated and preinoculated, is blended specifically for regional use, and may be planted in spring or fall.
- *Imperial Alfa Rack* is good for upland soils, hilltops, and hillsides. It provides up to thirty percent protein year-round and it also lasts up to five years after a single planting. It's preinoculated, specifically blended for regional use, and may be planted in spring or fall.
- *Imperial No-Plow* is for hard-to-reach areas or for landowners without farming equipment. It offers up to thirty-six percent protein, is an annual, and can be planted in spring or fall.
- *Imperial Power Plant* grows extremely fast, withstands heavy grazing pressure, and produces the high protein level that bucks need during the spring/summer antler-growing season. It contains a soybean and lablab and it can be planted through May in the South and as late as the end of June in the North. It lasts until the first frost, and even then it can provide some forage and good cover through the fall months.

Remington QuikShoots. Remington QuikShoots products are formulated to grow fast and provide a high-protein nutritional diet for deer, turkey, and upland species. QuikShoots blends can be planted in spring or fall and they have been tested in many areas of the United States.

- *Big Buck Blend* consists of sweet rapeseed and clover. Sweet rapeseed is an annual and the clover is a perennial.

- *Quik Clover Blend* consists of an annual rye and a special blend of perennial clover that will come back for three to five years.
- *No-Till Quik Clover* can be used to overseed an existing food plot.

Brier Ridge Wildlife. These seeds are blended by Olds Seed Solutions and seeds are available for both spring and fall planting.

- *Bucks Banquet* contains a mix of clover, leafy rapeseed, turnips, and chicory. Once cooler temperatures arrive, the mix becomes sweeter and more succulent.
- *Horn Honey Clover* contains a formula of high-quality clover seeds.
- *EZII Gro* is a low-maintenance perennial mix formulated to grow in areas where ideal preparation of the soil is difficult to achieve.
- *Rut N Ready* is a quick-grow mix designed to attract bucks during the rut. The mix contains high-energy turnips, rapeseed, and chicory.
- *Gobbler Gourmet* is a high-protein food source for turkeys. It attracts insects for spring and summer feedings and contains alfalfa, red clover, creeping red fescue, bird's-foot trefoil, Kentucky bluegrass, and white clover.
- *Rooster Relish* is a high-protein food source for pheasants. It contains buckwheat, early corn, black oil sunflower, early grain sorghum, and Japanese millet.
- *Fur & Feather Cover Mix* is formulated to provide deer, turkey, pheasants, and other wildlife with concealment cover. The mix contains early bird millet, early grain sorghum, huntsman millet, white proso millet, and Japanese millet.

Buck Busters. Buck Busters mixes have been field-tested under hunting conditions using a variety of soil types.

- *Fall Seed Mix* is very high in protein, with eighty-one percent digestible nutrients, and it includes three species of grass, winter peas, a mixture of brassicas, and crimson clover. Seed it from September 1 through October 15 in the Southeast, sixty days before the average frost date.
- *Summer Seed Mix* contains iron clay peas, forage soybeans, and hot weather corn varieties. Plant it after the Fall Seed Mix (above) has gone to seed, usually in late May or early June.

Tecomate. Tecomate's Food Plot System includes a variety of seeds.

- *Lablab* is a good warm-season source of protein (up to forty percent), antlergrowing phosphorus, and other key ingredients. Basically a super cowpea, it is nutritious, drought-resistant, and productive, averaging over 12,000 pounds/acre. Plant it in spring or, for bowhunting plots, in early fall.

- *Lablab Plus* contains lablab, ebony pea (a fast-growing vining pea relished by deer), and other big-seeded peas. Plant in spring or, for bowhunting plots, in early fall.
- *Outback Legume Mix* is formulated for the Gulf and Atlantic Coast states. It will not withstand severe winters and consists of coated and preinoculated warm-season sub-tropical legumes.
- *HomePlace Wildlife Mix* with Wildflowers attracts butterflies, songbirds, hummingbirds, turkeys, deer, and other wildlife to backyards. It includes coated and preinoculated white and red clovers and a variety of wildflowers.
- *Chicory* contains a mix of chicory varieties that is fast-growing, high in protein, and drought-resistant. It produces lots of high-protein forage and lasts for three to five years. Frost-seeding works well, and the mix is good for fall hunting plots.
- *Longbeard Turkey Mix* contains coated and preinoculated white and red clovers, yellow sweet blossom clover, bird's-foot trefoil, vetch, and chicory. It's an ideal mix for small woodland food plots. Plant it in the fall in the South; in April through July in the Midwest; or frost-seed.
- *Monster Mix* contains coated and preinoculated premium white and red clovers and chicory and provides deer with food year-round. It can be planted in spring or fall.
- *Ultra Forage Mix* contains legumes (peas, vetch, and clovers), chicory, and brassicas, and includes a hybrid rape/turnip that provides both forage and a highly edible bulb. Plant it in fall.
- *Max-Attract 50/50* contains fifty percent peas, clovers, vetch and chicory and fifty percent premium grain. A forty-pound bag plants an acre. Plant it in the fall to attract bucks.

PlotSpike. PlotSpike, from Regan & Massey, Inc., offers a wide range of food-plot products and blends that contain no fillers and have no chemical coatings.

- *Chufas* are great for turkey food plots. They grow in a wide range of climatic and soil conditions. A warm-season perennial, chufa produces subterranean tubers that look like nuts and have high levels of protein and carbohydrates. Plant in spring.

- *Spring Wildlife Mix* is a blend of warm-season plants that attract a variety of wildlife including turkeys, deer, and hogs. If you plant it in spring after the last frost date, it will provide food until frost in the fall.

Premium Fall Blend contains perennial prairie bromegrass, which is very palatable and long-lasting and provides high protein. It also contains forage rape, forage clovers, and a chicory.

- *Clover Blend* contains clovers and chicory. Chicory adds longevity to food plots during drought conditions.
- *Quick Stand Mix* is blended for quick establishment in varying conditions and can be planted with minimum tillage.
- *New Zealand Blend* contains forage rape and kale and is high in phosphorus and calcium. Combined with the clovers in the blend, it provides winter-long grazing and reseeding.

Wildlife Nutritional Systems of Texas. This company produces a wide range of food-plot mixes and blends.

- *Blue Magic Winter Legume* does well in the southeastern United States on dry upland soils.
- *Winter Deluxe* is a traditional oat, wheat, and rye blend. It also contains triticale, a wheat/cereal rye cross that is very winter-hardy and more palatable and disease-resistant than winter wheat.
- *Fall Blend* contains oats, triticale, Austrian winter peas, clover, and alfalfa. It is very easy to work with and does well on most soils.
- *Big Buck Xcellerator* is an economy version of the Fall Blend. It contains triticale, Austrian winter peas, and tyfon, a turnip/cabbage cross.
- *No-Till Buck Buster* is a fall-winter blend for those without access to farming equipment.
- *Spring-Summer Blend* consists of "Mr. Whitetail" Peas, browntop and pearl millet, white milo, and sunflowers. It is very fast maturing and can be double-cropped for spring planting in the North.
- *Upland Deluxe* is a mix for those whose primary concern is turkey, quail, doves, and other upland game birds. It consists of

gamebird peas, browntop millet, white milo, hegari, and sunflowers and produces cover and a large seed crop. Plant it in the spring.

- *Alfagraze* is a perennial true grazing-type alfalfa developed in the United States specifically for food plots. Bred to tolerate the heavy grazing that is common in food plots, it may stay green year-round in some parts of the country.
- *No-Till Mr. Whitetail Peas* provide an easy method of establishing wildlife food plots. Simply seed the peas and then cut the grass and weeds. The peas create a spring/summer and early fall food plot for deer, quail, and turkeys.

Antler King Trophy. Antler King Trophy has a wide range of products.

- *Fall/Winter/Spring Food Plot Blend* continues to grow and provide more than twenty percent protein during the winter when other plants die or become dormant.
- *Mini-Max Food Plot Mix* is a mix of perennial and annual seeds that have been proven to thrive in minimum tillage and the lower pH of more acidic soils. This mixture contains seeds for both northern and southern plots.
- *Trophy Clover Mix* is a blend of four clovers designed to provide high-quality forage for deer.

Schuster Farms. Schuster Farms has a wide range of products including lablab, an annual warm-season forage legume.

- *Spring Mega Mass* is a mix of fourteen subtropical legumes that complement each other throughout the growing season.
- *Trophy Forage Mix* consists of a unique blend of triticale, wheat, oats and rye varieties developed specifically for forage production.
- *Forage Plus* is a mixture of wheat, oats, triticale, rye and winter peas. The winter peas add protein.
- *Double Deer Mix* is a mixture of oats, wheat, triticale, rye, alfalfa, clover, peas, and vetch that provides excellent protein and energy.
- *Fall Premium Perennial Mix* combines seven popular clovers with three alfalfas, six aggressive medics or leguminous herbs and one mystery ingredient.
- *Fall Mega Mass Mix* has all the nutritional ingredients of the Fall Premium Perennial Mix plus with oats, wheat, triticale, and rye.

Pennington. Pennington is one of the largest producers of food plot and wild game seeds. Check Pennington's website for their extensive listing of seeds and wild game products.

If you don't purchase a seed blend, you may wish to plant a variety of annual and perennial seeds of different species, which provides wildlife with different foods at different times of the year. Since some plants are more drought-resistant than others, planting a variety will also help to ensure that some food is available even during dry weather. I plant several food plots in this manner, keeping the annuals and perennials in separate strips. Many companies offer food-plot mixes that include rapeseed, brassicas, clovers, cowpeas, turnips, triticale, lablab, and others.

Regardless of the seed you choose, proper planting is important for success. Carefully follow the manufacturer's planting instructions. Some seeds are planted in the spring, some in the fall, and some can be planted in either spring or fall. Some seeds are "no-till," which means they can be broadcast without thoroughly working the soil. You must, however, kill competing vegetation before you broadcast the seeds.

Seeds for Wildlife Other Than Deer

Although most of the products and information available today are for creating food plots to attract white-tailed deer, combination plots can attract other wildlife as well. Created properly, food plots can be magnets for wild turkeys, quail, rabbits, and numerous non-game species. The secret is to use plants that provide different types of seeds and forage that can be used by different types of wildlife. Since deer are opportunists, they will also appreciate these combination plots.

For example, clovers and alfalfa used for deer food plots also provide great bugging areas for turkey poults and young quail because they are attractive to many insects. And because these plants are relatively short, they provide easy forage for small ground-feeding birds. Clovers and alfalfa can be planted in combination with other plants such as annual grains or warm-season grasses to provide cover, grains, and other forage.

One example of combination seed packages is the Mossy Oak BioLogic Turkey T.O.P. or Turkey Optimization Program. It allows you to create adult bird forage areas, poult bugging areas, and grain as food for both. A thirteen-pound bag plants one acre. The seeds come in separate bags and are planted in strips, with the poult bugging areas on the outside, adult forage areas inside the bugging area, and the grain production seed in the center. The instructions with the seeds also suggest creating a dusting area as turkeys dust frequently and will return daily to the same location to dust if the conditions are favorable. You can create a dusting area in or near an existing food plot by frequent tilling or raking with a sturdy garden rake to maintain an area of loose soil. Dusting sites should be shaded near midday and be near a food source turkeys like. An ideal location is at the edge of a food plot beneath overhanging tree limbs.

You can create your own combinations, but it's important to match seeds to specific areas and soils. For the most part different species should be kept separate rather than mixing them together in one large planting because different seeds require different planting techniques. Some require merely pressing into the soil, while others require a slight covering. When you're planning your seed plot, consider that taller, spreading plants can prevent shorter plants from getting enough sunlight.

Strip Cropping Combination Food Plots

One of my favorite tactics is to strip-crop food plots. The more edge you create, the better the plot will be for all types of wildlife, and strip crops create a lot of edge. You'll need a plot of one acre or more; several of my plots are three to four acres. I prefer to create plots that are linear and that wind through cover such as timber rather than to create a large square block. One plan that has worked well is to place a strip of clover such as Whitetail Institute Imperial clover or Whitetail Institute Alfa Rack, which combines clover and alfalfa, along the outside of the plot. Then plant a strip of grain, such as milo combined with soybeans, next to the clover. Throw in some browntop millet as an added bird attractant as a separate strip. Deer will eat the soybeans to the ground, while the milo provides grain for turkeys and quail. Next leave a strip equal in size to the grain strip as open ground. This strip will grow up in ragweeds and other quail and dove foods with open ground beneath for ground-foraging birds. The clover is a perennial and will stay for several years. The grains are annuals and must be planted each year. You may wish to try alternating the open strip and grain strip each year to provide more versatility in the food plot. Or you could plant the third strip to green winter wheat or oats in fall for deer and turkeys and plant milo and soybeans in the spring. I've used all of these combinations with good results.

The strips should be twelve to fifteen feet wide and you can create as many as you desire. You can even add a wider sunflower strip on the outside of one edge to create your own mini-dove field. Or try a dual patch—seed one with spring-planted species and another with fall-planted species to create more versatility in attracting whitetails and other wildlife. Both Mossy Oak BioLogic and Pennington offer numerous spring, summer and fall planted seeds.

Since the annual portions of these fields require replanting each year, you may also wish to consider planting an entire field in perennials. An extremely effective combination is a plot with clover around the outside and warm-season grasses in the center. The native warm-season grasses such as switchgrass, big bluestem, little bluestem, and Indian grass are all bunchgrasses. This means that they grow tall and upright in bunches with bare ground between the plants, making them especially good at providing both cover and food for quail and turkey poults, pheasants, and small mammals such as rabbits. The seed is relished by wildlife, while the foliage is an excellent deer food. Deer like to bed in the grass because it hides them well, and they especially like it if the area is close to another food source such as clover. The best method of managing warm-season grasses is with regular prescribed burns. The clover strip around the outside of the plot provides a natural firebreak, but it must be wide enough to create a good break; twenty feet should be adequate in most instances. (See Prescribed Burns.)

One simple method of establishing clover food plots is to over-seed with a legume in late winter.

A very simple combination green browse food plot was recommended by the Missouri Department of Conservation over twenty years ago: Uniformly seed a one-acre plot with a half bushel of wheat and two pounds of orchard grass in early September. At the same time or in early winter, overseed half of the plot with two pounds each of ladino clover and red clover. During the following January through March, overseed the other half with ten pounds of Korean or Summit lespedeza. This legume plot with a thin stand of grass provides forage for turkeys, deer, and rabbits and also attracts insects for turkeys and quail. The lespedeza provides seeds for quail and green forage for other wildlife during summer periods when clovers may become dormant.

Disking Plots for Annual Weeds

One of the simplest methods of creating food plots, especially for turkeys and upland game such as quail, is disking and leaving the land fallow. This allows annual weeds, especially ragweed, to sprout. These weeds supply seeds and food without the expense of planting. One method recommended by many experts is to maintain a portion of a food plot in disked, fallow land each year. Then rotate, planting it the following year and disking another portion of the plot.

When to Plant

Timing is important when seeding food plots. Check the seeding times on the product bag and follow the times recommended for your state or zone explicitly—or check with your county extension office about recommended planting times. I like to make both spring and fall plantings—or rather cool- and warm-season plantings. Doing this provides not only new growth and more food year-round, but also a better variety of foods and some drought protection. For example, I plant annuals such as corn, soybeans, milo, and peas in the spring, and I also plant perennials such as clover or alfalfa in the spring. The perennials can also be planted in the fall, and in some areas that's best because there's less chance of drought and less weed competition for the tiny seeds. In the fall I plant winter wheat or oats. In late fall I broadcast ladino clover over the wheat or oats, creating a perennial plot for the next year.

Advantages of fall planting. About the time the summer ends and cooler nights begin, many hunters start to think about the coming hunting season. I'm in the same mode at this time of year, but I'm also thinking about food plots. I've discovered that fall-planted food plots offer some advantages over spring-

One of the easiest food plots is made by simply disking the land and then leaving it fallow. The natural weeds that sprout will provide seeds and food for turkeys and quail. COURTESY PLOTMASTER

planted ones, especially plots of clovers and alfalfa such as Imperial Whitetail Clover and Alfa Rack.

Lack of weed competition. Clovers planted in the spring have to sprout and then come up through weeds and grasses that are usually already in place and have a jump on the tiny clovers. Even in a tilled food plot, they'll have some competition from sprouting grasses and weeds unless the weeds and grasses have been sprayed with an herbicide prior to planting.

Attractive food during hunting season. If all conditions are right, fall-planted food plots can offer extremely succulent new growth during the hunting season, attracting deer more readily than old growth food plots do.

Early spring establishment. Come next spring, plants in fall-seeded plot are already established, making it easier for them to compete with the weeds and grasses. This means that the fall-planted food plots will be at their prime during the spring months when whitetails need nutrition the most. Does in the last stages of pregnancy and bucks beginning antler regrowth require good nutrition at this time when there are few other food sources, and a new clover food plot is an excellent source.

More time available. It's easier for me to find time to plant in the fall than in the spring. Regardless of whether you're doing the work yourself or hiring someone to do it, spring is a busy time, especially for farmers trying to get their crops planted. Fall is not as busy for most farmers, so it's usually easier to hire someone or find time yourself.

More convenient liming. In order for clover and alfalfa to use the nutrients in the soil or the fertilizer you apply, the pH of the soil must be correct—between 6.5 and 7.0. It takes time for lime to break down in the soil and work with the nutrients, and the time required depends on the type of lime applied. Because of the large amounts of lime often needed, commercial application is usually the best choice. This means applying the lime when food-plot areas are dry so the trucks won't get stuck. Late spring or early summer are good times for lime application for fall-planted food plots, and these are usually times when applicators and lime suppliers are not overbooked.

To have a successful fall food plot, several factors are important:

A good seedbed.

Getting seed in early. You must plant fairly early in the fall if you want to hunt over the food plot the same year.

Adequate rainfall. Regardless of whether you plant in the spring or fall, moisture is needed, and in many parts of the country rainfall is sometimes scarce in the fall.

It is important to choose seeds that are adapted to fall planting. Imperial Whitetail Clover is an excellent choice and my all-around favorite. I have about twenty-five acres planted in eight food plots scattered around my farm. The clover stays green here in Missouri until mid-winter, when the ground freezes solid. Then the clover dies back, but it's one of the first plants to rejuvenate in the spring. By late February or early March I see lots of wild turkeys in the clover plots, and their bright green droppings indicate where they're getting

Turkeys love clover and are often the first animals to frequent a food plot once it greens up in the spring.

their first good nutrition after the long winter.

Unfortunately, not all soils are suited for clover. Well-drained, drier soils, particularly on hilltops, often do best with an alfalfa product. Most alfalfas tend to die back more quickly than clover and a very hard frost can brown out some species. In fact, this can happen just before or during deer season, which can quickly change deer feeding patterns and, of course, hunting success. Whitetail Institute's Alfa Rack, however, is a clover-type creeping alfalfa blend that is less susceptible to early browning. Alfa Rack is a special blend suited to different parts of the country. It requires only six weeks to germinate, giving it a good start before winter hits.

Planting Steps

Following the proper planting procedure is important for both fall and spring food plots. Before you plant, take soil samples to determine the pH and soil nutrients your soil needs. Ideally, this should be done in late spring so that you can determine lime requirements and apply the lime. Then break ground. In central Missouri, where I live, I like to till the plot in early August. If I'm creating a new plot on an old field or a similar area, I first apply an herbicide to kill existing vegetation. Once the vegetation has died back, I apply fertilizer and till the soil.

Clover seed is hardy enough to plant it in the fall, winter, or spring. It will lie dormant until it germinates with the warmer temperatures. COURTESY LEE HOARD

You can plant clover seed anytime from fall through winter and into spring. The seed will lie dormant until there is enough sunlight and warmth to start to germinate it. If you're late getting your fall seed planted, plant winter wheat as a cover crop. This prevents soil erosion and keeps the tiny clover seeds from being washed away in case of heavy rains before germination takes place in the spring. I cleared a four-acre food plot two years ago and by the time I planted, it was too late for fall germination. Wary of erosion due to the bulldozer's action, I planted triticale over the plot as a cover crop. The triticale came up fine and I had lots of deer in the plot throughout the winter. The next spring I had one of the best Whitetail Imperial Clover germination successes I've experienced, and that plot, at this time, is one of my most productive.

One unusual but popular technique farmers and ranchers use to upgrade grass pastures is overseeding with legumes in the winter. I've done this with good success using ladino and red clovers to create food plots. Doing this also offers some of the advantages of fall planting. Overseeding is a great way to add quality nutrition to a grass stand without actually breaking ground. Imperial No-Plow is an excellent choice for this method and it is also extremely easy to use. For a total clover plot, rather than a mix of grasses and clovers, you must kill the weeds and grasses with herbicide before planting and then broadcast in mid to late winter. This works best when the ground is frozen. As the ground thaws in the spring, the seeds seep into the cracks in the soil created by the temperature changes. An even better idea is to spread the seed on snow cover. Not only does this allow you to see your seeds so you can keep your coverage even, but as the snow melts in the spring the seeds are distributed down into minute cracks in the soil. This method also provides an initial nitrogen boost. Legumes such as clovers add nitrogen to the soil. As the snow melts, it also adds nitrogen to the soil.

CHAPTER 4

PREPARING THE LAND

Regardless of whether you've just purchased land or already own land you wish to improve for wildlife, it's important to assess the habitat. If you're lucky, you have open land ready to till and plant. In many instances, however, land will need to be cleared. This can involve brush-hogging old fields, dozer work, or prescribed burns.

CREATING NEW FOOD PLOTS

The first step in creating a new food plot is clearing the land of existing vegetation. Depending on the vegetation, the location of the plot and other factors, a variety of methods can be used, and over the years, I've used a number of them.

Old fields or overgrown clearings can make excellent food plot locations. These can sometimes be cleared of saplings and brush with a chainsaw and brushcutter.

Old fields. If you have an old field in early to mid-succession growth, or the early growth of woody plants (brush), sometimes simply disking with a big offset disk will work, though you may have to make several passes in order to completely break down the brush and weeds. If you don't have the equipment you can often hire someone to do it. You can also use a brush-hog or large rotary-blade mower to cut the vegetation down to ground level. Swisher and Weekend Warrior make a heavy-duty model that can be pulled behind an ATV for clearing brush and weeds.

One old overgrown field on our property had hardwood saplings up to four inches in diameter, and in many places they were so thick you couldn't squeeze between them. I spent most of my noon breaks and several Saturdays over the course of a winter cutting out these sprouts using a lightweight chain saw and a brushcutter. A brushcutter resembles a string trimmer except that it has a large saw blade on the end instead of string. The better units, such as those from Stihl, have a shoulder harness and handlebar controls for easier and safer use. You can cut saplings up to three inches in diameter with some of the larger units. I cut the bottom sections of the larger saplings into firewood and hand-carried all the tops to the clearing edges, where I stacked them into big brush piles. A hen turkey used one of the brush piles as a nest site the first year. When cutting the saplings, it's important to cut as close to the ground as possible without damaging the saw blade or chainsaw chain.

On several similar old fields I first cut only the larger saplings, again making sure they were cut extremely close to the ground, and then used a front bucket blade on my tractor to push them into brush piles along the edge of the woods.

Regardless of the method you use, the saplings and/or weeds and brush will come back almost immediately, so you also must spray the area with an herbicide to kill the weeds, brush, and stumps. If you're planting a crop such as Imperial Whitetail Clover or Alfa Rack, it's important to totally kill the vegetation; otherwise the tiny seeds won't be able to germinate. Herbicides are available for controlling different types of vegetation. Saplings should be treated with a stump-killer as soon as they are cut. Brush should be treated with a brush-killer before clearing. Weeds should also be treated before disking. Tall, vigorous weeds should be mowed down and allowed to regrow, as most herbicides work best on new growth.

Instead of using herbicides, you may choose to do a prescribed burn several years in a row. This will eventually kill the saplings as well as any cedars that may have invaded the pasture. You may also use a combination of herbicides and prescribed burns.

Woodlands. Clearing woodlands is different from clearing old fields because you may need to use heavy equipment such as a bulldozer.

Large-scale clearing is best done with a dozer, although it's more expensive. Bulldozing also tends to remove valuable topsoil.

Unfortunately, dozers also remove some of the topsoil, the amount depending on the soil type, vegetation type, and the skill of the operator. Dozer-cleared areas also tend to end up fairly rough unless you pay extra for the operator to smooth the area. You can also smooth the land with a tractor and a box scraper. The final smoothness of the land also depends on how many rocks are exposed during the dozing process. The smoother the cleared land, the easier it is to plant and the better your chances will be for a productive food plot. Land clearing with a bulldozer is not cheap, but in some cases it's the only option.

Choose a place to pile trees removed during the dozing and cut the smaller ones into firewood. Logging-value trees can be removed first and sold, or you can cut them into lumber using a portable bandsaw mill such as the TimberKing. These one-person sawmills are easy enough for even a beginner

An alternative in some instances is to kill back trees by frilling or girdling, then spraying them with an herbicide.

to use, and with them you can create building materials for home, hunting camp, or sale. If you have quite a lot of valuable-timber land, a mill can make your land investment more viable.

An alternative to full-scale bulldozing—or any type of mechanical land clearing—is to kill vegetation, including trees, with herbicide and then use a no-till seed such as Whitetail Institute's No-Plow. You won't end up with as pretty a food plot, but the animals won't care. First kill trees so light can reach the ground. Cut completely through the bark and completely around the tree; this is called "frilling" or girdling. An axe, hatchet or chainsaw can be used, but you must cut completely through the bark and cambium. Squirt a tree- or stump-killer herbicide into the frill. You can also use soil-active herbicides to

A Plotmaster equipped with disks and shovels will handle most heavy-duty tilling tasks.
COURTESY LEE HOWARD

kill trees growing close together in groves. It will take at least a season for the trees to die. Once they're dead, spray the remaining vegetation with an herbicide such as Habitat Release from BASF, which will control low-quality hardwood brush, allowing beneficial plants such as lespedeza, blackberry, and partridge pea to thrive. Or you may prefer to kill all vegetation using Round-up herbicide. Once the plants turn brown, use a prescribed burn or a brush-hog to remove the vegetative cover. You can pull a Plotmaster fitted with a blade or rake behind an ATV to help clear the ground for more even planting. However you clear the land, you must lime and fertilize the exposed soil just as for any other type of food-plot preparation. Then scatter the seed, such as Imperial No-Plow, over the ground just before a rain. No-Plow is preinoculated, so you don't have to hassle with inoculation to provide a better germination. No-Plow seeds are also coated to protect them from moisture and prevent false germination.

CHAPTER 5

IMPROVING SOIL FERTILITY

We are all—plants, animals, and humans—products of the soil. The fertility or richness of the soil determines the health, vigor, reproduction, and often the size of animals and plants growing on it. The kinds and amounts of food elements like iron, phosphorus, calcium, and nitrogen in the soil are important for plants and in turn for animals. For instance, if there isn't enough lime or phosphorus in the soil to grow teeth and bones and to make good blood, wildlife populations will be small and unhealthy. And if the soil provides too few vitamins, animals will be diseased and fail to bear healthy young. This is one of the reasons deer, turkeys, and other wildlife in agricultural areas are, on average, bigger and have larger populations than those in low-fertility areas. Good farmers are good land stewards,

The soil must be tested for fertility and pH. Soil samples are collected and then sent to soil-testing laboratories.

and they know that in order to grow food they must keep the soil fertile. If you manage a hunting area for wildlife, your first priority is to produce high-quality and nutritious foods. You can do this by creating annual or perennial food plots, providing pasture forage such as warm-season grasses, or increasing the vigor of native plants. In all of these methods, adding fertility or maintaining soil fertility is of major importance.

Testing Soil

Before you start amending the soil, you must determine its present fertility and acidity (pH) with a soil test.

Collecting equipment. You can obtain soil sample containers and information sheets from university extension offices and some fertilizer dealers. If you don't have "official" containers, put the soil in clean plastic receptacles such as cottage cheese or margarine tubs with tight lids. You'll also need a clean plastic pail for collecting and mixing the samples (galvanized pails contain zinc that can contaminate samples). Special soil tubes or augers are available for collecting samples, but a garden spade will do for heavy or rocky soils and a garden trowel for lighter soils.

Soil classification map. Although it's not strictly necessary, a soil classification map is a good resource. They are available from university extension or Soil and Water Conservation Service offices. The map should include information about differences in soil type, slope, and past crop growth and yield in your area, if any. The soils in each area you test should be the same type or classification: color, slope, surface texture, internal drainage, and past erosion.

Information to provide soil testers. It's important to provide soil testers with a cropping history (if there is one) and a future planting plan, as different types of plants require different amounts of fertilizer for optimum growth.

The collection process. Each soil sample should be composed of twelve to fifteen subsamples representing no more than forty acres for level, uniform fields or five acres for hilly, rolling land or nonuniform land.

- *Collecting a subsample.* To collect a subsample, first scrape off surface litter—leaves, twigs, grass, etc. If you use a soil tube or soil auger, collect one core at each subsampling site. If you use a spade or garden trowel, dig one V-shaped hole six to eight inches deep and remove a half-inch-thick slice from a smooth side of the hole. Then trim off sides of the slice with a trowel or similar tool, leaving a one-inch-wide strip.

- *Combining the subsamples.* Repeat this procedure in twelve to fifteen places, and put the subsamples together in a clean pail. After collecting all of the subsamples for one soil sample, break up the clods and thoroughly mix all of the samples together in the pail. Transfer a portion of this mixed soil into the soil sample container. Discard the remainder of the sample material and clean the equipment before taking samples in another area.
- *Labeling.* Label each container with your name, address, and a sample number corresponding to the number on a soil sample information sheet or simply in a notebook: for example, Field A, Sample 1. Prepare a map or sketch of your food plots and their locations on your property. Label the areas sampled and include the date of the samples. This will provide you with a record of your soil testing history that will be useful in the future. Fields or plots should be retested every four years to determine availability of lime and fertilizers. (Most areas will require annual maintenance fertilizers, but the lime will not need replenishing as often.)
- *Send your samples and receive results.* Take or send the samples to your local university extension center where they'll be tested for a nominal fee, usually $5 to $15 each. In ten to fourteen days, sooner if you provide an e-mail address, you'll receive results indicating how much, if any, lime is required to adjust the pH to the plants you intend to plant, how much organic matter is in the soil, and how much "fertilizer" you need to "bring the soil up to test." Fertilizer can include potassium, calcium, phosphorus, magnesium, sulfur, zinc, manganese, iron, copper, and nitrogen or nitrate.
- *Using the results.* Going by the test results, a fertilizer dealer can mix the precise amounts of minerals you'll need to apply. "Plants are basically nutrient transfer agents," explained Grant Woods. "In other words, if the nutrients aren't in the soil, the plant can't give the nutrients to the wildlife. Start with the soil, do a soil test, and follow the recommendations."

The proper pH is extremely important for a productive food plot. In many instances lime will have to be added.

Fertilizer and lime are your biggest expenses in planting food plots. A soil test can determine the exact amounts of each you need so you won't apply too much or too little.

Liming

The proper soil pH is required for successful production of grain, forage, and fiber crops, including food plots. The pH is a measure of the soil's acidity or alkalinity and is indicated by a number from 0 (the most acidic) to 14 (the most alkaline). Neutral is considered to be 7.0, and this the desired pH for growing most plants. Soils in most parts of the country are acidic—below neutral. Soils in woodlands that have been cleared for food plots are normally quite acidic due to the collection of leaf matter. Although plant foods such as calcium, phosphorus, and potassium may exist in the soil, they may not be available to plants because they're bonded to the acidic elements. Plants such as clovers and other legumes, which fix nitrogen in the soil, do not do as well in acidic soils because the nitrogen-fixing Rhizobium bacteria is inhibited in such soils. In most instances you'll be throwing away time and money on fertilizer, tilling, and seeding if you don't first adjust the soil pH with lime.

The benefits of adequate liming include:

- Decreased soil acidity
- Improved effectiveness of fertilizers
- Immobilization of elements toxic to plant growth
- Provision of calcium and magnesium
- Encouragement of desirable micro-organic activities
- Improved structure and tilth of some soils

Lime is produced by grinding limestone into small particles. Limestone available for agricultural uses varies greatly in quality depending on the fineness of the grind as well as the chemical composition. The neutralizing effectiveness of ground limestone varies from source to source. Limestone producers perform laboratory analyses of their products that provide the pounds of effective neutralizing material (ENM) in a ton of lime. This value is obtained by tests for calcium carbonate equivalent and fineness of grind. To determine the amount of lime to apply to a plot you've had soil-tested, divide the suggested ENM on the soil test report by the ENM in one ton of the limestone product you plan to use.

It is usually considered impractical to apply less than two tons of limestone per acre. If less than two tons per acre is recommended in the soil test results, it's best to apply two tons or wait a year or two until the pH drops. If large amounts of limestone are required, it's best to make two applications. For instance, if eight tons per acre are required, apply four tons the first year and four tons the second. Many food-plot seed producers suggest applying seven tons per acre as a general guideline.

The soil's response to liming occurs over time. It's important to thoroughly mix the limestone in with the soil layer so the lime is available to the plant roots. Limestone for food plots is usually applied by bulk truck or by bulk wagons pulled behind tractors. Agricultural lime is also available bagged, but it's quite costly to buy it this way even for small food plots. Limestone should be applied at least six months—and preferably a year or more—prior to planting or seeding crops that are sensitive to excess acidity, including most of the clovers and alfalfas.

Different soils require different frequencies of lime application. Sandy soils require more frequent applications, while clay soils require less frequent applications but need a larger initial application. It is a good idea to take a soil sample every three to four years to determine soil acidity.

Understanding Soil and Fertilizers

Fertile soil is necessary to grow plants, including food plots. The fertility of the soil is dependent on several factors:

- An adequate supply of plant-food elements.
- Sufficient moisture to supply the plant foods to the plant roots.
- Enough warmth to encourage plant growth.
- Air to supply oxygen.

The organic makeup of the soil is important. Friable or loose soils with organic matter, including soil bacteria and necessary minerals, grow the best food plots. Heavy clay-type soils are not as productive because the plant foods and moisture can't get to the roots as well and less oxygen is available. Sandy, dry soils don't hold moisture and also allow plant foods to escape from roots. Different types of soils hold different amounts of nutrient elements or cations: [calcium, magnesium, potassium, and the non-nutrient cation hydrogen (neutralizable acidity)]. The capacity of soils to hold these cations is called Cation Exchange Capacity, and it depends on the kind and amount of clay and organic matter in the soil.

Positively charged ions, known as cations, are held by negative charges of the soil colloids (clay and organic matter). Plant roots exchange hydrogen ions (acidity) for calcium, magnesium, and potassium ions. The resulting effect on the soil is depletion of the nutrient ions and increasing acidity. Soil treatment becomes a matter of replacing neutralizable acidity with a satisfactory balance of calcium, magnesium, and potassium.

Desirable values of exchangeable cations for Balanced Soil Saturation are seventy-five percent calcium, ten percent magnesium and 1.7 to 5.3 percent potassium for heavy clays to sandy soils respectively. Saturating of calcium may vary from fifty to seventy-five percent and magnesium from six to thirty-five percent providing that the sum has a minimum value of eighty-five percent saturation.

Providing Soil Nutrients

As explained above, plants need nutrients, including magnesium, phosphorus, potassium and nitrogen.

Magnesium. Magnesium (Mg) is found in limestone. The amount of magnesium carbonate per ton of ground limestone varies.

Phosphorus. Phosphorus (P) has many important functions in plants, the primary one being the storage and transfer of energy through the plant. Adenosine diphosphate (ADP) and adenosine triphosphate (ATP) are high-energy phosphate compounds that control most processes in plants including photosynthesis, respiration, protein and nucleic acid synthesis, and nutrient transport through the cell walls. Phosphorus is essential for seed production, promotes increased root growth, produces healthy growth, and encourages good fruit development. Rock phosphate is the most common source of phosphorus.

Potassium. The most common source of potassium (K) is muriate of potash, derived from large deposits of potassium chloride salts found in the southwestern deserts of the United States. Potassium is needed for the manufacture of carbohydrates—sugars and starches. Potassium also increases resistance to disease and produces strong plant cell walls and stems.

Nitrogen. Nitrogen (N) is necessary to convert the sun's light into energy through photosynthesis. Plants also use it to form amino acids, the building blocks of protein. Protein is needed by wildlife such as deer, and the availability of plant-usable nitrogen often determines the quantity and quality of forage available. Plants with dark green leaves have a high or proper amount of nitrogen availability. Slow-growing, stunted plants or those with yellowing leaves show a nitrogen deficiency.

Each plant has different nitrogen requirements. Corn, for instance, has one of the highest nitrogen requirements, needing as much as 215 pounds of nitrogen per acre to produce high yields. Legumes, such as clovers, alfalfas, and soybeans, "fix" or add nitrogen to the soil. Rhizobium bacteria form nodules on the roots of the plants and then take nitrogen from the atmosphere and make it available to the plant. Although these plants add some nitrogen to the soil, they often use most of it as well. If they are planted with other plants, you may need to add some nitrogen to the soil. Applying nitrogen early in the season, however, allows grasses and weeds to get a jump on the legumes. Apply the nitrogen in the late spring or, better yet, late summer.

Fertilizer Types

Commercial fertilizer is available in several forms. You can have a dealer custom mix several fertilizers according to the recommendations of your soil test. This method allows you to tailor the nutrients to your needs and not have

The proper amount and types of fertilizer must also be added to the soil, according to the plants to be grown.

too much of one or the other. You can also buy premixed fertilizers in bulk or in bags. In this case a numerical code on the bag indicates the amounts of each element. The first figure is the percent of N (nitrogen), the second is the percent of P (phosphate or phosphorus), and the third is the percent of K (potash or potassium). For example, one hundred pounds of 8-31-16 has eight pounds of nitrogen, thirty-two pounds of phosphate, and sixteen pounds of potash. The rest is inert materials.

Fertilizers may be spread using a variety of means. Larger areas require bulk application.

Some commercial fertilizers are designed specifically for food plots:

Scotts Food Plot Fertilizer has an analysis of 32-3-10. Most university extension personnel recommend that when you use an ordinary commercial fertilizer, you make one feeding at planting time and follow it with additional feedings, especially nitrogen, throughout the year, a costly and time-consuming task. Scotts Food Plot Fertilizer, however, is a timed-release product, so you can feed just once a year—at planting time. Each forty-three-pound bag feeds one-fourth acre.

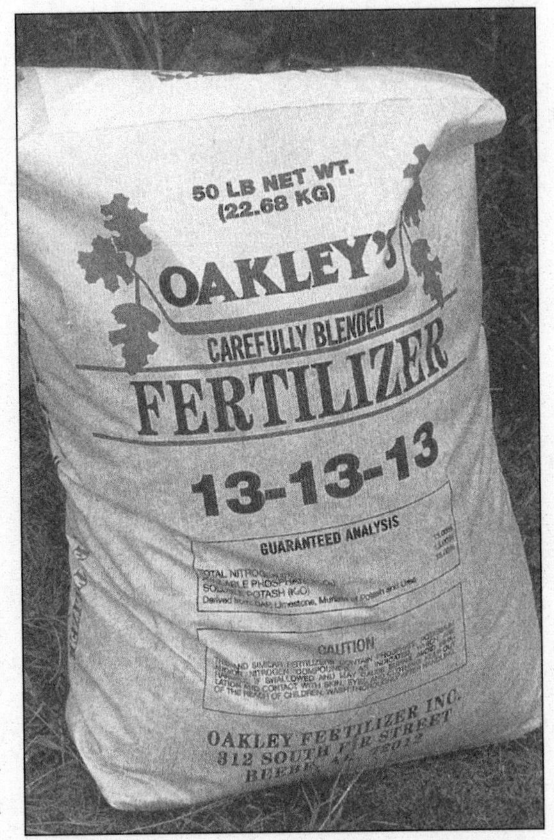

Fertilizer may be purchased by the bag. The numbers represent the percentage of each component.

Pennington Penngreen is an all-purpose fertilizer that can be used on most soils for many food-plot crops. Also available from Pennington is their Wild Game Food Plot Fertilizer (15-5-10).

Biologic's pHFertilizer is a basic 10-10-10 fertilizer with finely ground lime as the filler component, allowing you to fertilize and lime at the same time. The formula also contains the appropriate quantity of micronutrients such as boron and sulfur, based on regional soil types.

Tip: After spreading fertilizer, make sure to thoroughly clean the spreader as well as the ATV or tractor, hosing off fertilizer dust and particles. Fertilizer is extremely corrosive and can rust metal parts if not washed away.

CHAPTER 6

PREPARING THE SOIL AND PLANTING

Killing unwanted vegetation and tilling the soil are important steps to take before you plant. Different seeds require different planting methods.

UNDERSTANDING AND USING HERBICIDES

Herbicides clear land of trees and brush and eliminate weeds from food plots. They are designed to kill plants, and they can be nonselective, which means they kill all plant matter they touch, or selective, which means that used properly they will kill one type of plant but not injure another. When using selective herbicides it's extremely important to follow instructions regarding weather conditions, concentration, and methods of application. Too little of the herbicide may not kill unwanted plants; too much may injure desired plants. Apply them only as directed and only to approved plants. Most selective herbicides break down in the soil, and the rate of decomposition is determined by rainfall and temperature. Often

Some grasses, such as fescue, are extremely invasive and will come back even after tillage. These grasses should be killed back before planting.

spring-applied herbicides are inactive by summer, so you may need to reapply them to keep weeds from germinating.

Herbicides are categorized as *preplant, preemergence,* or *postemergence.*

Preplant herbicides are applied before a crop is planted.

Preemergence herbicides kill weeds while the seeds of the weed are germinating, but not after growing plants have developed green coloring.

Postemergence herbicides kill plants after they are green and growing actively. They can be used to kill annual weeds that have emerged from the soil as well as some perennial weeds.

Understanding herbicide labels is important. Each herbicide has four names: common name (shorthand name of the chemicals in the herbicide), chemical name (the herbicide's chemical makeup), trade name (producer or manufacturer), and brand name (the name given the product by the producer).

Applying Herbicides

The easiest way to apply herbicides to food plots is to spray them. Hand-held sprayers can be used to spot-spray weeds and brush. Sprinkler cans or squirt bottles can be used to treat stumps and for tree frilling. Tractor- or ATV-mounted or pulled tank sprayers are best for treating large areas of brush and saplings. Boom sprayers can be used to treat grass and fields. Make sure to follow spray rates for the ingredients used. Granular herbicides may be broadcast using fertilizer spreaders.

It is almost impossible to remove all herbicide traces from a sprayer after use. It's best to designate one sprayer as a herbicide sprayer and mark it as such. Don't use that sprayer to apply insecticides or fungicides to desirable crops or plants.

Choosing the Right Herbicide

Many herbicides are available, and it is important to match the herbicide to the plants you wish to kill.

Killing grasses and weeds. Common herbicides for control of early spring weeds in no-till fields of corn, soybeans, grain sorghum, and small grains are 2-4D, atrazine, dicamba, glyphosate, paraquat, and metribuzin.

Some grasses, such as fescue, are so invasive that they won't allow other plants to compete, yet they offer no value to wildlife. Even after tillage they,

Brush and small saplings that have been cleared from the area will also resprout unless treated with an herbicide.

along with many weed plants, will emerge in food plots. Round-up (glyphosate) is commonly used to kill vegetation in no-till operations when creating food plots.

Two herbicides have been developed primarily for use in wildlife management. Habitat Release and Plateau Eco-Pak, available from BASF, work by affecting an enzyme found only in plants, not in humans, mammals, birds, insects, or fish. One application of Habitat Release controls low-quality hardwood brush so you can establish and maintain wildlife openings and food plots.

Killing brush and trees.

- *Cut stubble and stumps.* Use a mixture of eight ounces of Crossbow, eight ounces of Tordon 22k and two gallons of diesel fuel. Sprinkle or pour the solution onto the cuts immediately after cutting; it can be applied year-round. Spray the cambium layer of the cut (the layer between the bark and the wood), the bark of the stump, and all exposed roots. Overspray may kill what it contacts, so try to keep the formula only on the stump and exposed roots.
- *Brush and saplings at full leaf stage from April through July.* Use a foliar spray: one hundred gallons of water, two quarts of Crossbow, three pints of Tordon 22k, and two quarts of a surfactant such as Silkon. Spray. This general mix works well on oaks,

locusts, sassafras, elms, and cedars, but there will be some regrowth on liberty elm and multiflora rose. It costs about a dollar a gallon and it takes about one hundred gallons per acre for a thick stand of brush.

- *Larger trees.* Use a streamline basal application: thirty-two ounces of Crossbow, thirty-two ounces of Tordon 22k, and two gallons of diesel fuel. Spray a five- to six-inch band around the tree six to twelve inches above the ground.

If plants you've treated with a foliar application start to grow again, wait two growing seasons before you treat them again so you'll have enough surface area to spray. If you use either basal spray or foliar application, don't cut the brush or tree until the next growing season. Cutting a tree too quickly after spraying interrupts the translocation of the spray into the root system and results in regrowth.

Herbicide Precautions

- Read the safety precautions on the herbicide label.
- Keep new or unused herbicides in their original containers and keep them locked out of the reach of children.
- Don't allow herbicides to contact your skin or eyes.
- Don't smoke while using herbicides.
- Wear a rubber apron, heavy rubber gloves—not thin surgical ones, and goggles when mixing herbicides. Your face (especially your eyes), hands, forearms, stomach, and thighs are the areas of your body that absorb chemicals most rapidly.
- Record in writing what you are mixing or spraying and have that information and a product label with you. In case of an emergency you will then have the phone number of the company and a list of antidotes readily accessible. If you are unconscious, the person who finds you will be able to use this information.
- After using herbicides, wash your skin thoroughly.
- Don't pour unused herbicides down the drain or into streams, irrigation channels, or drainage ditches.

Preparing the Seedbed

The better the seedbed is prepared, the better are your chances for a successful food plot. The amount and type of ground-breaking required depends on the prior use of the land. The area may require disking, or it may need to be deeply plowed. Thin, rocky, hard, and clay soils require more

A properly prepared seedbed produces the most productive food plots. The soil must be well broken and smooth.

preparation than deep, loam-type soils. Regardless of the type of soil, break the ground and work and rework it until all dirt clods are broken up and the seedbed is smooth.

Seeds such as wheat, corn, and soybeans will grow in rougher seedbeds than the smaller seeds like the clovers. Clover seeds are tiny, and in a poorly prepared plot with large dirt clods and clumps of mud or clay, germination will be spotty. My granddad was a good farmer, and he also prided himself on his garden. He used a team of big red Missouri mules to plow the ground. Then he hitched them to a disk and disked the plot to further break up the clods and smooth the plot. Finally, he used a harrow to drag the plot and break up the soil clumps even more. It was a lot of work, but when he finished you could shoot marbles across the flat, smooth ground. The horses these days are a lot easier to use and more powerful, but the technique needed for success is the same. The smoother you can get the ground, the better the germination rate and the better your chances for success. I take my granddad's process one step further when I'm planting tiny seeds and pull a big field roller behind my tractor or pull my Plotmaster with its field roller behind my ATV to compact the soil and make it smooth and flat.

Despite the information above, sometimes you can plant food plots on bare ground without plowing or tilling. Use herbicides to kill the vegetation followed by a prescribed burn to remove the dead, dry vegetation. Then simply throw the seeds right on the burned ground, and roll them in place.

SEEDING

Tiny clover seeds require not only a fine seedbed but a firm one as well. Success with these legumes requires some steps you don't have to take with larger seeds. After the seedbed is prepared, use a roller or cultipacker to firm the soil. Then broadcast or drop the clover seed directly on top of the soil. Follow with the roller or cultipacker to lightly press the seed into the soil. "If I get ten complaints a year about food-plot success, five will be because people planted their clover seeds too deep," said Steve Scott of Whitetail Institute.

"I suggest planting just before or during a rain," says Grant Woods. "The raindrops will press the seeds in just right."

Other seeds require planting deeper and covering to the proper depth. Seeding can be done with a broadcast seeder that is either hand-, tractor-, ATV- or truck-powered, or the seed can be dropped into place with planters or a grain drill. Small grains, such as wheat, oats, and rye, can be broadcast or drilled using

Be sure to know the proper depth for the type of seed you are planting. Nothing is more discouraging than not seeing your plant emerge because you planted the seeds too deep in the soil.

a grain drill or a planter grain drill attachment. Milo, soybeans, and other beans can be broadcast or drilled. Corn is usually planted with row planters. I like to set broadcast seeders to throw about half the amount called for and then make more passes in different directions to ensure even coverage. Then I make sure to cover them to the recommended depth. When you use a drill, the seed depth is set by the operator of the machine. Light disking is also a simple and fast way to plant these seeds. Disk, then broadcast and disk again to cover the seeds.

Seeding can be broadcast by hand or with a broadcast seeder.

The biggest problem for many food-plot managers is buying the equipment to manage their plots. Large farming equipment is expensive, even if you buy it used. You may be able to hire a nearby farmer to do your ground-breaking. If you do, consider planting a perennial legume such as clover, because clover plots don't have to be redone each year. Many chores, such as broadcasting seed, can be done by hand with a crank-type broadcaster. Today, however, a wide range of food plot management tools are available for use with ATVs and lawn-and-garden tractors. (See Chapter 8, Food Plot Tools.)

CHAPTER 7

FOOD PLOT CARE AND MAINTENANCE

Even with some of the best food plot seed in the world, and with years of farming experience, I've had failed food plots. When you plant a food plot, you're farming, and like farmers worldwide, you're at the mercy of Mother Nature. But Mother Nature isn't the only cause of food plot crop failure, so alleviate the farming problems you can control, then hope or pray to avoid those you can't control. Food plots can be planted just about anywhere crops can be grown without irrigation. The amount of rainfall is the prime factor in having a good crop, but even the more arid areas can produce if managed properly. Although much of the following has been covered in the preceding chapters, here are some of the main reasons for food-plot failures.

IMPROPER AMOUNTS OR TYPES OF FERTILIZER

The most common reason for crop failure is lack of lime and/or the proper amount and type of fertilizer. A soil test will determine lime and fertilizer requirements. In many parts of the country application of lime is required to adjust the pH of the soil and to allow the plants to take up the fertilizers. It takes time for lime to be absorbed into the soil, so it must be applied at the appropriate time.

POOR LOCATION

Some soils and locations just don't produce as well as others. As with other farm crops, rich, loamy soil is better than thin, rocky soil.

Inadequate Ground Preparation

If you're creating a food plot in an existing field of grass and weeds, especially fescue, the vegetation should first be killed with an herbicide. Even with heavy disking, not all the fescue will be killed, and it will quickly come back along with a variety of weeds.

Improper Planting Techniques

A common mistake is burying tiny seeds, such as clover and alfalfa, too deeply. The tiny seeds probably won't germinate, but even if they do, they may not have enough energy to push through the heavy layer of soil. These small seeds should just be pressed into the topsoil or barely covered. A field roller or a drag can be used for this chore after seeding. A simple drag can be made with of a couple of pieces of pipe and a section of chain-link fencing. In a pinch I've sometimes even cut down a ten-foot cedar tree and used it as a drag. If rain is forecast, simply broadcast the seed on a smooth seedbed and let the raindrops incorporate the seed into the soil.

Inadequate Weed Control

You wouldn't dream of planting a new lawn and then watching the weeds take it over. Any time ground is broken, weeds will come up, especially in a newly planted plot. By late summer the annual weeds, and probably some grasses, will have grown tall and overshaded the clovers. In clover food plots, mow just often enough to cut the weeds and clip the top inch or two of the clover as well. This will cut back the annual weed growth and rejuvenate the clover.

Poor Seed Quality

Make sure you use good-quality, viable seeds. Seeds such as those from Whitetail Institute are not only top quality but also preinoculated. If you don't inoculate seed such as clover, you can expect only a fifty to seventy-five percent germination rate. Inoculation can be done by hand, but the job is messy and not very accurate. If possible, plant fresh seed. Check the seed date on the seed bag label. If you store seed for the next season—or longer—germination may be reduced, and pests such as insects may destroy the seed. If you must store seed, use an air- and insect-tight container. To check for viability of old seeds, place twenty-five seeds in a damp—but not soggy—paper towel. Roll the towel up, fasten it with rubber bands, place it in a plastic container, and cover loosely. Place the container in a warm place, out of drafts. Wait until about a week after the average seed germination time (this information is on the seed label) and then open the packet, count the sprouted seeds, and multiply by four. This will give the percentage of germination. Germination rates average eighty-five to ninety percent or better on quality, fresh seeds. You can expect twenty-five to fifty percent germination on year-old seeds, and even less on older seeds.

Low Moisture

If you've paid attention to the points above, you've done all you can, but your crop can still fail if there isn't enough rain. This happened to me several years ago. A drought hit Missouri just about spring planting time. I went ahead with the planting and hoped for rain, but it never came and the plot failed. A couple of years later the same thing happened. This time, however, the drought ended in late July and, guess what, my spring-planted seed came up with the first rain! So don't give up too early on dry-season food plots. The success rate wasn't as high as I had expected, but I still had a food plot. After about a year of growth, plots are less susceptible to drought. I have about a dozen food plots and like to keep them in rotation, planting or rejuvenating two or three each year. This way I hedge my bets in case Mother Nature causes a crop failure one year.

Food plots can be planted under dry conditions, but don't expect them to be as successful as those planted under normal rainfall conditions. The biggest problem is that if soil is plowed or disked very deeply, it loses a lot of moisture through evaporation from the newly exposed soil. If you must plow to break the ground, do so before the rainy season, even if that's some time before you will be planting, and then use minimum tillage at the time of planting to prevent deep soil evaporation.

Successful food plots are not guaranteed, but following the tactics above can prevent many food plot failures.

Food Plot Maintenance

Food plots don't require a lot of maintenance, but they do require some.

Mowing. Some food-plot plants, such as clovers and alfalfa, should be clipped or mowed (fairly high) in late summer. A mower behind your ATV may be able to handle this. Mowing is especially important on clover, because in the summer and early fall, weeds and grasses often grow above it, cutting down on its productivity. Light clipping or mowing at a fairly high level removes the weed and grass tops and also the tops of the clover, stimulating growth. The clover responds with lush new growth that deer relish. The result: deer are more attracted to the plot.

Fertilizing. After the first year's initial fertilizing and planting, apply a light amount of fertilizer to maintain perennial food plots. Follow the maintenance schedule recommendations that came with your soil test.

Perennial clover food plots need maintenance. Light clipping in late summer removes weed growth and rejuvenates the clovers.

Keep records. It's a good idea to keep a record of each plot. Include planting times, seed and fertilizer information, and the occurrences (times and numbers) of wildlife observation or harvest.

Check productivity. To find out how productive certain varieties are and how much forage is being grazed, try placing utilization cages (simple wire baskets that prevent animals from grazing) in food plots to keep deer from grazing specific areas.

Year-Round Schedule

Question: If you stick some seeds in the ground in April, will you have big bucks in November? *Answer:* Not necessarily. Food plots are a year-round affair. Like any type of agricultural pursuit, they require maintenance as well as proper planning, with some chores best done during each season. To provide a continuous source of food, consider planting in both the spring and fall. Here are some food plot chores and the best seasons in which to do them. (Keep in mind that seasonal variations across the country allow people in the South to plant earlier than those in the North.)

Winter

- *Take stock of what you've accomplished.*
- *Set goals for next year's food plots.* Examine existing food plots both early in the winter and late in the winter and determine whether you need to replant or create new plots in the spring. Legumes and some other perennial plots may need to be replanted if they are more than four or five years old. Decide where you wish to plant new food plots. I like to use a piece of tracing paper over an aerial photo to mark boundaries, possible food plots, etc. Remove the tracing paper, pencil in the information for the food plots or areas, and make a photocopy. You can then mark on and use this plot plan, carrying it in the field with you without losing or marking up your valuable aerial photo.
- *Start a record book with each food plot listed.* Enter dates of plantings, types of plantings, when fertilized, and when limed, as well as records of animals harvested or observed on the plots.
- *Take a soil sample* if your soil isn't frozen solid or covered with snow.
- *Use the test to determine lime requirements* and have the lime spread while the ground is solid, before the spring thaws and rains make it impossible for heavy lime trucks to get to your plots. It takes time for lime to take effect, and many plants, such as legumes like clovers, require lime to be spread six months or more before planting.
- *Fertilize existing food plots* in late winter.
- *Perform prescribed burns* on food plots, wood lots, and warm-season grass areas, or to help remove unwanted grasses for establishing food plots. This is best done in late winter or early spring.
- *Clear areas for food* plots in late winter.
- *Widen old logging, farm, or ranch roads.*
- *Clear old fields* that have grown up in brush and poor-quality forage to create new food plots.
- *Overseed with legumes* while the ground is frozen.
- *Establish minerals and salt licks* for deer by late winter.

Spring

- *Add fertilizer* to bring the soil in new plots up to test.
- *Break ground and start planting.* Legumes, such as clovers, and alfalfas can be planted in the spring, but they often do best in the fall when they don't have as much competition from weeds and grasses. Spring, however, is the best time to plant annuals—peas, millet, sorghum, and wildlife grazing mixes such as warm-season food-plot mixes and cool-season annual mixtures. You can, of course, plant clovers in the spring if you don't think you'll get a chance to do it in the fall.
- *Perform prescribed burns.* In some parts of the country early spring is the best time for burns.
- *Fertilize existing food plots* and/or existing vines and trees.
- *Plant dove fields* by the end of May in most areas.
- *Mow perennial food plots* such as the clovers to control weeds if needed.
- *Plant food plots for ducks.*
- *Lime for fall plots.* If you're planning fall food plots, get soil tests at this time and apply lime according to the soil test recommendations.

Summer

- *Place utilization cages* in your food plots to determine the amount of grazing early in the summer months.
- *Lightly clip perennial food plots* such as clovers through the summer months to remove weeds that may shade them and other more desirable plants. Ordinarily you should mow high to avoid disturbing the perennials, but if you're managing for turkeys, mow lower to promote regrowth of the perennials and to provide turkeys with brood-rearing areas.
- *Mow, rake, and bale dove fields* in late summer if it's legal in your area.
- *Start breaking ground and preparing for fall planting* by late summer, usually around the first of August.

Fall

- *Plant fall and winter annual food plots* such as winter wheat, oats, ryegrass, and fall seed blends. You can also plant cool-season perennials such as clovers and alfalfas.

Spring is an ideal time to fertilize existing food plots, shrubs, and trees. COURTESY U.S. FISH & WILDLIFE SERVICE

- *Apply maintenance fertilizer* to existing perennial food plots.
- *Spray herbicide if you're creating food plots on old grass fields,* particularly fescue, and then in the spring you can burn the dead vegetation.
- *Create firebreaks* around fields to be burned in the spring by disking, brush-hogging, or mowing.

By using a year-round food-plot schedule you can spread out the workload and provide better quality food plots for all types of wildlife.

CHAPTER 8

FOOD PLOT TOOLS

Many landowners or lease hunters don't have the farming equipment needed to create food plots, but the ATV, a tool you might already have in your hunting camp or garage, may be able to help you prepare, plant, and maintain food plots. A number of accessories are available for creating and maintaining food plots with ATVs or lawn-and-garden tractors, and of course you can use the same implements to till, fertilize, and plant your garden and care for your lawn. I've tested a number of them on my own property.

SPRAYERS

ATV rack-mounted or pull-behind sprayers are very useful for creating food plots and killing brush as well as for managing timber and performing prescribed

A wide range of ATV accessories make putting out food plots quick and easy. A boom sprayer allows for killing back fescue and weeds.

burns. Sprayers are available with a single hose for spot spraying and detachable booms for field spraying. Models are available from Outland Sports (API), Cabela's, Cycle Country, and Bass Pro. Cycle Country's AG Commercial ATV Sprayer, with a twenty-five-gallon tank, mounts directly onto an ATV's rear rack and produces 3.0 gpm at 45 psi. Standard equipment includes an eighteen-inch poly gun that allows for spot spraying directly from the ATV seat. A boom can be added to make short work of spraying larger areas.

BROADCASTERS

Broadcasters can be used with both fertilizer and seed and can also be used to broadcast supplemental corn or other feed for wildlife in areas where this is legal. They're easily mounted on ATV racks and switched off and on as needed. Models are available from Cabela's, Moultrie, API, Cycle Country, and Bass Pro.

Cycle Country 100-pound-capacity spreader. This spreader mounts to either the front or rear rack of an ATV, and both the spread pattern and flow rate are adjustable.

Moultrie Feeders ATVSdx Spreader. This spreader attaches to the rear of an ATV and is easily removed. It's easy to use, with a handy feed control bar and on-off switch.

Spreaders are designed to hang on ATV racks, and filled with fertilizer they can be very heavy. I constructed a simple support bar that runs from the bottom of the spreader and rests on my Kawasaki hitch. It goes on and comes off with

Broadcast seeders can be used to broadcast fertilizer or seeds.

wing nuts and helps to distribute the weight of the full spreader. I also welded a bracket on a disk to hold the spreader, which allows me to spread seed and/or fertilizer and cover it at the same time.

TILLAGE IMPLEMENTS

Tillage implements are available for both single hitches and three-point hitches. Single-hitch accessories, the most common, use the existing ATV hitch and are fairly easily pulled. Single-hitch multipurpose tools are available.

Swisher Quadivator

The Quadivator works with a garden tractor or an ATV. It has electric depth control and can convert to a rake, tiller, irrigator, lawn roller, chemical applicator,

The Swisher Quadivator is a great multi-purpose unit. It can be converted to a rake, tiller, irrigator, lawn roller, aerator, disk, or plow.

aerator, disk, and plow. The tandem disk installs with a single bolt. Optional equipment includes the Auto Dump Box, which attaches to the frame and allows easy dumping from the operator's seat; a box scraper/leveler that can be used for landscaping and other leveling chores; hilling moldboards; a potato/vegetable digger; a lawn irrigation plow; a barbed wire dispenser; a lawn roller; and a lawn chemical applicator.

Woods-N-Water Plotmaster

The Plotmaster is designed for planting food plots with an ATV or small tractor. It comes with a single-hitch pull-type tongue and a three-point hitch. The unit can be used with larger lawn tractors with three-point hitch capacity or smaller farm tractors of forty horsepower or less.

The basic unit weighs five hundred pounds and is solidly built, but I pulled

The Plotmaster performs nine different functions using a variety of attachments. It allows one-pass planting, tilling, dropping seed, and cultipacking.

A wide range of accessories is also available for the Plotmaster, including a blade, rake scraper, and compactor.

it fairly easily with a Kawasaki Prairie 300 4WD ATV. A bigger ATV would make the chore even easier. My friend, longtime deer hunter Peter Fiduccia also uses a Plotmaster on his farm. He pulls it with a Honda Rincon 649cc 4WD ATV. Fiduccia says pulling the Plotmaster with a Rincon is easy—it's a real workhorse of a machine. If you put a broadcast spreader on the front of your ATV and pull the Plotmaster, you can take everything you need to the food plot with one implement.

The Plotmaster is good for single-pass planting of small seed such as clovers, but it can also easily be adjusted to plant soybeans, peas, corn, and other larger seeds. It includes a variety of attachments for different functions.

- *Harrow.* The disk harrow can be adjusted to three positions to change the aggressiveness of the cut.
- *Sweep plow.* Smoothes the soil.
- *Chisel plow.* Cuts through compacted soil.
- *Cultivator.*

- *Drag.* Levels the soil
- *Seeder.* A drop-type seeder is built into the unit.
- *Cultipacker.* Presses the seed into the soil and firms up the bed.
- *Optional equipment* includes lawn aerator, a rake, a scraper blade, and a compactor.

Cycle Country Products

Cycle Country offers small three-point-hitch tillage implements designed for larger 4WD ATVs. Here are some features :

- *Quick change implements.* With the mounting hardware and hitch in place, the implement can be attached or removed quickly.
- *Lift.* An electromechanical screw-driven lift raises or lowers the accessories as needed.
- *Lockout hitch.* The three-point hitch locks out the rear suspension of the ATV itself.
- *Float.* A built-in float added to the hitch provides smooth operation of the implements while they're in use.
- *Attachments available:*
  - a *single-bottom moldboard plow* with a maximum tillage width of ten inches, an adjustable depth gauge wheel, and a fourteen-inch beam clearance. Use this plow for initial groundbreaking, and then smooth the seedbed, breaking up clods into finer pieces, using a disk harrow.
  - a *harrow* with eight fourteen-inch blades and five adjustable angle positions with a maximum tillage width of forty-six inches.
  - a *furrower* that creates one deep furrow and can be used to create firebreaks as well as plant furrows for trees, useful if you're putting in a tree plantation. It can also be used in the garden to dig furrows to plant potatoes and then to dig the harvest at the end of the season.
  - a *seven-tooth spring-tine cultivator* that has seven-inch shovels with a half-inch overlap, a maximum tillage width of forty-seven inches, and an optional row-crop kit that cultivates row crops such as soybeans, milo, and other annual food-plot seeds.
  - a *rear blade* with a maximum scrape width of sixty inches straight and forty-four inches angled.

Cycle Country's three-point hitch attaches to ATVs and uses an electro-mechanical screw to hold accessories.

Cycle Country also produces a one-row planter that's useful for planting annual food plots of corn, soybeans, and milo as well as helping you do your gardening chores.

Cycle Country's Spring tooth cultivator helps simplify cultivating chores.

Bombardier Products Bombardier produces a line of professional-grade implements that operate off a multi-tool carrier with an electric actuator. Eight attachments are available, including a disk harrow covering a fifty-inch span with eight disks; a fifty-inch cultivator with seven shovel-head tines; a roller with a fifty-inch drum; a dethatcher/harrow with thirty-five spring-tooth tines spread over fifty inches; a fifty-inch lawn aerator; a forty-two-inch sweeper with industrial bristles; a fifty-inch box scraper; and a cargo box that can be used with or without any of the other attachments.

Cycle Country's one-row planter makes planting corn easy.

Wish list: I wish someone would manufacturer a lightweight cultipacker; it's the single best implement for precisely positioning of fine seeds such as clovers in the soil. A good alternative is the Sears Craftsman lawn roller that can be pulled behind an ATV to compress the fine seed into the soil and smooth out the seedbed.

The Weekend Warrior tandem disk is a heavy-duty, sixty-four-inch model with electric actuator for ease of use.

Mowers

Cycle Country Rough Cut

The Cycle Country pull-behind-your-ATV Rough Cut mower can handle tall grass, thick weeds, and brush. The Rough Cut has a forty-eight-inch cutting width and a quick-pin offset hitch so you can mow to the left or right of or directly behind the ATV. This allows you to mow close to fences, under trees, or even along water while keeping the ATV on firm ground. Legumes such as clovers do best if lightly clipped during the summer, which also removes high-growth weed competition, and the Rough Cut is a good choice for that chore.

The Weekend Warrior universal tow frame also accepts a wide variety of accessories.

Swisher Trailcutters

The forty-four-inch Trailcutter mower is designed for clearing brush from overgrown fields, roadsides, fence lines, and trails. It has a 10.5-horsepower engine and heavy-duty swinging blades that cut most brush up to 1.5 inches in diameter. It has a single-point height adjustment and single-pin hitch,

Swisher has a wide variety of mowers for ATVs including their Trailcutter that will cut brush up to 1½ inches in diameter.

allowing it to fit a wide variety of ATV hitch heights, and a universal articulating hitch with an offset tow bar allows you to mow off to one side of the ATV. The low height of the ATV and the offset mower also make it a good choice for mowing pond dam faces.

Swisher also makes a fifty-inch Trailcutter with the choice of a 12.5-horsepower Briggs and Stratton I/C engine or a 13-horsepower Tecumseh OHV engine. It has a 12-volt electric key start with throttle control and has a floating cutting deck with rear discharge so that you don't scalp uneven ground. The company also manufactures two pull-behind finish-cut mowers for those who prefer a finer end result.

Maxxis Extra Traction Tires

The Maxxis Scout front tire features chevron-shaped lugs that dig in and provide excellent steering control and traction in all terrain types. The Maxxis Hunter rear tires are specially designed for today's larger 4x4s. The specially designed tread bars are spaced to shed loose dirt without losing traction, making them ideal for ground-breaking chores.

ATV Safety Rules

General Rules

Wear a helmet. This not only protects you in the event of an accident, but also protects you against injury from stones or other objects thrown while mowing or doing other chores.

Wear goggles or safety glasses mowing and doing other ground preparation chores.

Clear the area of objects such as rocks, logs, or wire that could be thrown by the blades when you're mowing.

Disengage the power to the mower and stop the engine when transporting the mower or when it's not in use.

Turn off the mower blades when you're not mowing. Never leave the mower unattended with the engine running and blades turning.

Rules for Working on Slopes and Around Obstacles

Take extra care in going up, down, and sideways on slopes. Spreaders, sprayers, mowers, and other attachments can create heavy loads on the ATV. Working on slopes is a major factor in loss-of-control and tip-over ATV accidents, and such accidents can cause severe injury or death. If you feel uneasy on a slope, don't work on it. If you do work on a slope, pull the equipment up and down the slope—never across.

Watch out for obstacles. Pulling implements can create tip-over situations, especially if the pulled implement contacts obstacles such as stumps, roots, or heavy rocks. Pull implements only in low gear, and if you encounter a stump or rock, back up, lift the implement, and go around or over the obstacle. If possible, remove all obstacles such as rocks and tree limbs before you start. Watch for holes, ruts, and bumps, and be especially wary in tall grass and weeds, which can hide obstacles and uneven terrain that can overturn an ATV, especially one that's pulling equipment.

Tractors and Attachments

I began life as a farmer—and I'm still one. Although I write for a living, a farm has always been a part of my life. As a result, much of my experience with food plots and other wildlife management has been with farming equipment—I've always owned a farm tractor and attachments. In the past few years I've also experimented with almost every ATV attachment available, and many of them work fine, but there are some chores ATVs just don't do quite as well as farm implements. In fact, the ideal situation is to have an ATV for some chores and a full-size tractor and appropriate attachments for others. When you move up to the big-boy toys, however, the prices escalate. A good 4x4 ATV and the equipment needed for food plots costs around $10,000.

Choosing a Tractor

A new fifty-horsepower tractor, on the other hand, which would be the minimum acceptable size, can run over $50,000. But bargains are still to be found in used equipment. (It would be great if automobiles lasted as long as most tractors do!) Older used tractors can be found at implement dealers and farm auctions. Don't expect them to be cheap, though. It's a good idea to check out some pricing before you hit an auction. Although at least a fifty-

For the larger habitat management projects, a full-size farm tractor may be the best choice.

horsepower tractor should be your first choice, you may find that a newer twenty- to forty-horsepower model with four-wheel drive is suitable for many chores. Try to pick a model with wide front wheels. The older tricycle-type tractors are less manageable and can be unsafe in some of the rougher areas you may need to reach to work on food plots.

Tractors are either gas- or diesel-powered. Gas is less finicky but more expensive to operate. The more powerful models are mostly diesel. Try to get one with power steering—it's well worth the added cost. You may also opt for a cab, but the extra height can be a problem when you're trying to get under overhanging limbs bordering food plots. Tractors with an up-exhaust can also be a problem around overhanging limbs. I've torn off a number of exhaust pipes over the years trying to negotiate wooded areas.

Purchase a tractor with a three-point hitch and power take-off. The hitch allows you to easily raise and lower implements with a lever from the tractor seat. The power take-off provides power to an implement such as a brush-hog. Several years ago I ran an old 9N Ford with direct power right into a pond during brush-hogging. Even though I had my foot on the clutch and brake, the force of the brush-hog kept the tractor moving. Newer models feature independent power take-off. With independent power, the forward or reverse motion of the tractor is independent of the power take-off.

Tractor Attachment

Front Bucket

The single best attachment you can have for a wildlife-management tractor is a front bucket. You can transport and hoist tree stands, lift deer for dressing, shove brush piles, cart firewood, and even use it to haul seed and fertilizer to the food plot.

Ground-Breaking Implements

- *Moldboard plows* were used for many years, and they can't be beat for really breaking the soil deeply. But they require a lot of power, and you have to smooth the ground afterward with another implement and another pass.
- *Offset disks.* This is probably the best single implement for food plots. Buy the heaviest one you can use with your tractor, soil types, and conditions. Double tandem disks are better than single

blades. An offset disk breaks the ground, rolls it together, and also breaks up any trash, such as last year's cornstalks or weeds. Look at both new and used. You can expect to pay at least $1,000 to $2,000 for an older eight- to twelve-foot model.

- *Monroe-Tufline* offers a full line of disks ranging from four-foot models to twenty-four-foot hydraulic-fold models for 220-plus horsepower tractors. The Bio series of three-point-hitch tractor mount disks is available in 6.4-inch, 6.89-inch and eight-foot cutting widths. They offer eighteen- or twenty-inch notched disk blades with adjustable gang cutting angles. Optional equipment includes heavy-duty adjustable blade scrapers, aggressive super-notch disk blades for heavy ground cover disking, and smooth disk blades for fine finish requirements. Six- and eight-foot rear-mount drag harrows can also be mounted behind the disks for top-dressing applications.
- *BrushMaster* disks come in different widths and blade configurations. They feature adjustable gangs for different field applications—the steeper the angles of the gangs, the more aggressive the disking. An optional broadcast seeder with bracket and dragboard kit is designed to throw seeds directly behind the BrushMaster disk in the disked dirt and allows the dragboard to cover the seeds.
- *Great Plains* Discovator is available in several sizes, including the smallest—12/5—which combines disking, cultivating, harrowing, and herbicide incorporation in one pass.
- *Other disks.* Lightweight three-point disks are available at many farm supply houses. These will work, but they do require added weight. I use one such disk for a great deal of my ground work and constructed a wooden box of treated lumber that sits on top of the disk and holds concrete blocks as weights.

• *Harrows.* Many seeds, especially the tiny clovers and alfalfa, require an extremely fine seedbed. A harrow is the best choice for this, and many harrow models are available at farm and tractor supply stores.

- *Tillers.* If you don't have rocky ground or ground with a lot of tree roots, tractor-powered tillers work well. I have a four-foot-wide Long tiller with my Long tractor, but rocks, stumps, and roots can create real problems. The Till-Rite Rotary Tillers from BEFCO are available in five sizes to fit tractors ranging from sixteen to seventy horsepower. The built-in manual side-shift system allows you to operate the machine shifted to the right or centered to cover tractor tire tracks.
- *Fertilizer spreaders.* The BEFCO Turbo Hop three-point hitch power take-off fertilizer spreaders are available in a number of hopper sizes. Fertilizer literally eats metal, but the Turbo models are made with reinforced fiberglass hoppers and nylon/reinforced fiberglass spreader spouts for longer life. BEFCO also manufactures pull-behind models for smaller tractors. Monroe-Tufline offers the SBT series of spreaders—one thousand-pound tractor-mount and six hundred-pound ATV-mount.
- *Herbicide sprayers.* Three-point hitch, tractor-mounted, and pull-type models are available.
- *Seed drills.* Precision seeding, especially when planting light, fluffy seeds such as native grasses or tiny seeds such as clovers, is best done with a seed drill. Great Plains has a wide variety of both No-Till and Mini-Till drills that are great for planting food plots, CRP (Conservation Reserve Program) grasses and native warm-season grasses.
    - *No-Till Drill.* This six-foot, three-point is designed primarily for wildlife conservation, vineyard/orchard seeding, pasture renovation, roadside seeding, and mine reclamation. It is not an overseeder, but a primary seeder. The no-till method involves spraying herbicide, then seeding over the killed vegetation. The No-Till models feature 450-pound preloaded coulters to cut through the dead vegetation and a ground-driven metering system, which is the best way to seed warm-season or native prairie grasses.
    - *Min-Till Drill.* This precision seeder comes in five-, six-, seven- and ten-foot widths, is speficially designed to seed

Precise seeding of grasses and legumes is made easy with the Great Plains Mini-Drill. COURTESY GREAT PLAINS

into a prepared seedbed, and gives you the option of three seed boxes. The main seed box uses a fluted design to seed a variety of seeds, from peas to alfalfa. The small seed box uses a smaller version of the fluted feed cup to meter clover, alfalfa, and other small seeds. The native-grass box-picker wheel feed design is aggressive enough to feed fluffy native grass blends like big bluestem, Indian grass, and side oats grama.

- *Multiple-use attachments.* The Woods-N-Water Plotmaster four-foot model can be used to create food plots with ATVs or small tractors (less than forty-horsepower.) An eight-foot version is designed primarily for larger tractors and has the same basic features as the original, allowing for one-pass planting. You can disk, plow, cultivate, plant, and cover or cultipack in a single pass. You can also de-thatch, aerate, reseed, and cultipack in a single pass. The units come with a one-point hitch and a three-point hitch, so you can operate the Plotmaster with many types of machines. Each unit comes with a multi-use disk harrow, sweep plow, chisel plow, cultivator, electrical seeder, cultipacker, and drag. Other attachments are available, including a moldboard plow, turning plow, firebreak plow, rolling basket, aerator, compactor, rake, scrape blade, one-row planter, sprayer, and a grain drill.

- *Cutter/Mowers.* Rotary cutter/mowers can be used to maintain food plots and clear trails as well as to perform other chores. I keep several woodland clearings on our property in early succession growth by cutting back sprouts each summer. This keeps the ground cover down and allows more growth of forbs and other plants that deer relish. BEFCO makes rotary cutters to fit tractors from twenty to one hundred horsepower.

With a full-size tractor and the right implements, food-plot planting, maintenance, timber-clearing management and CRP creation or management is easy.

CHAPTER 9

ASSESSMENT AND MANAGEMENT PLAN

If you own or purchase property with timber, you'll have lots of deer, turkeys, and other wildlife, right? Not necessarily. The amount and type of wildlife will depend on the amount and type of woodlands as well as the health of your woodlands. These three factors are all "manageable." Properly managed woodlands can provide not only better timber production, but also opportunities for improving forest game conditions.

Timberlands and woodlands or woodlots also need proper management not only for wildlife habitat, but for producing wood products. In many instances the same practices apply for both.

For example, when we purchased our Missouri Ozarks farm back in the early '70s, approximately sixty-five percent of it was wooded. It was an old, worn-out dairy farm. Cattle had grazed the timber and the timber hadn't been harvested for a long time. We could see two hundred yards through the park-like woods. One of the reasons we bought it was to hunt deer, but the first year we hunted hard for a week and saw only one deer.

Our first step in improving the situation was to selectively log some areas and clear cut some others to create wildlife openings in the woods. Less than three years after our efforts started we began to see lots of deer. Another wooded property we purchased later was a different story. It hadn't been grazed for a number of years, but it had been logged off a few years before. The majority of the woods had grown up as "pole" timber—thick stands of trees that are all the same size. Again, we could see several hundred yards through the timber. With crowded crowns, the mast crop had continually been low and there was little wildlife usage. We also managed that property for wildlife primarily by selective thinning, and in the process produced a large amount of firewood.

Cathedral-like woodlands of mature forests have closed canopies that shade out the forest floor. This keeps other plants from growing, including vines, shrubs, young trees, forbs, and grasses, all important foods for most wildlife. In managing timberlands, the goal is to provide a diversity of woodland types and habitats, ranging from open areas for wild turkeys to thickets for deer. In many instances, a chainsaw may be the wildlife's best friend.

A number of practices can be used to improve woodlands for wildlife, or even create woodlands, including:

1. Improvement of existing timber stands
2. Leaving some den trees or creating den trees
3. Saving some fruit-bearing trees
4. Excluding livestock from timberland
5. Improving woods borders for wildlife
6. Retaining existing clearings
7. Renovating clearings
8. Creating new wildlife openings
9. Reforesting (planting)

Create a Management Plan on Paper

Don't just head out to the woods with your chainsaw and start cutting. First, make a management plan. The plan should consist of short-term goals, mid-range goals, and long-term goals. And don't just say, "I want to manage for wildlife." You need to be more specific. For instance, you may desire to have more wild turkeys on your property; more deer; more grouse. You must manage specifically for these purposes, although many practices benefit all wildlife. As with food plots, a topographical map and aerial photo are excellent tools for tabletop planning. Once you've used them to identify the land you want to manage, contact state or local forestry departments—or private foresters—for advice. These professionals can assist in evaluating your forest and your goals. They'll help you draw up a management plan specifically suited to your property and the types of wild game you wish to manage. Most state and federal foresters offer free advice, while private foresters are available for a fee.

Regardless of your specific management goals, the more diversification in your woods, the better your chances for greater wildlife populations. Try to provide all the requirements for wildlife in different parts of your woodlot or timberlands. These should include deer bedding areas; refuges; feeding areas; nesting areas for birds; bugging areas for turkeys and quail; foods such as hard mast, soft mast, forbs, grasses, and legumes; and brush. Maintain a cutting and maintenance schedule to keep uneven aged trees and to provide food and cover.

Take a Timber Cruise

After you've done your tabletop planning, make an inventory of your woodlands. The inventory should describe tree species, stand densities, composition, age, tree diameters, heights, quality, and growth rates. Before you begin this inventory, draw a map of the property. Outline property boundaries, forest boundaries, and other land uses, including croplands, pastures, and fields. Indicate roads, trails, and utility right-of-ways. Mark the locations of buildings, ponds, and streams. Soil maps can also be important in planning woodland management, because some species grow better in certain types of soils. Garner as much information as you can concerning past use of your woodlands. Have they been grazed? When were they logged? Have there been previous fires or serious pest problems? The

The first step is to make a timber cruise, or an assessment of the existing woodland types and sizes.

more information you can gather, the better you can make your management plan.

If you have small woodlots, you can assess the property and make a management plan fairly quickly. If you have larger timbered areas, it's essential to have a timber cruise made by a professional forester. The cruise determines the makeup of your forest—the age, numbers, and types of trees. On larger properties the timber cruise and resulting management plan may divide the woodland into separate areas depending on management goals and the makeup of the woodland. Each area, or stands, comprises two to forty acres of trees of similar species, ages, sizes, distributions and numbers of trees per acre. The stands are usually fairly easy to define, because roads, trails, fields, streams, and bluffs surrounding them act as natural boundaries.

You'll treat each stand as a specific management unit and follow practices fairly uniformly throughout a given stand. This allows you to tailor management practices to suit specific sites or timber types. By overlaying a diagram of the stands over a map or aerial photo, you can also get a better overall picture of your property and goals, making it easier to determine short-term, mid-term, and long-term goals and decide where you want to start.

Dividing the timberland into stands of similar site and tree characteristics also allows you to allocate money and chores to specific goals. With larger woodlands, it's frequently most practical to work on stands in rotation. For example, by thinning a different stand each year you provide diversity, which is good for attracting wildlife.

The following is a typical timber-cruise plan for one stand from our property. The cruise and plan were made by the Missouri Department of Conservation forester in our area.

FOREST STAND NUMBER 13, ACRES: 11

Soils & Topography: Ocie-Gatewood complex, 15-35%;
Goss gravelly silt loam, 3-8%

Management Objective(s): Oak timber and hard mast production

Vegetation:

Dominant Trees: Black oak, post oak, honey locust, and hedge

Plants & Shrubs: Black oak, elm, and a variety of perennial grasses and forbs

Description of Stand Condition: This stand has also seen some harvesting activity in the western third of the stand. This portion of the stand consists of poor- to fair-quality medium saw timber-size oak trees. The middle third consists of poor-quality post oak, and the eastern third consists of poor- to fair-quality black oak and hickory large pole timber.

Stand Management Needs/Treatment Recommendations: Conduct a precommercial thinning favoring the healthy single-stemmed white and black oak trees. The trees to be removed should be cut or girdled with a chainsaw. Trees can be girdled with an axe, but an herbicide such as Tordon RTU should be applied when an axe is used. All stems with a diameter greater than two inches and less than eight inches should be cut within six to twelve inches of the ground on the uphill side. Trees eight inches in diameter and greater may be girdled. The girdled chain-saw cuts must completely encircle the tree, sever the cambium layer, be one inch deep, and be below the lowest live limb. The cuts must connect! Hickory and honey locust trees should be girdled twice and the girdles must be at least two inches apart and one-half to one inch wide when made with a chainsaw and two inches wide when made with a hatchet or axe.

Another type of timber cruise is taken primarily for measuring sawlogs, and the resulting information is recorded on a tally sheet. This provides information on the number and types of trees per acre, the basal area, and the stand density. On small properties or in stands of extremely valuable timber such as walnut logs that could be used for veneer, individual trees are measured. On extremely large properties it is not practical to measure every tree. In this instance sample plots are measured following a grid pattern. The plots can vary in size from ⅕ to $\frac{1}{20}$ acre. A ⅕-acre plot is about 106 feet in diameter. The number of plots will vary with the acreage and type of timber, but at least one plot per acre should be sampled. To establish a grid, you'll need a compass or GPS. The diagram below show a typical grid pattern. Determining the

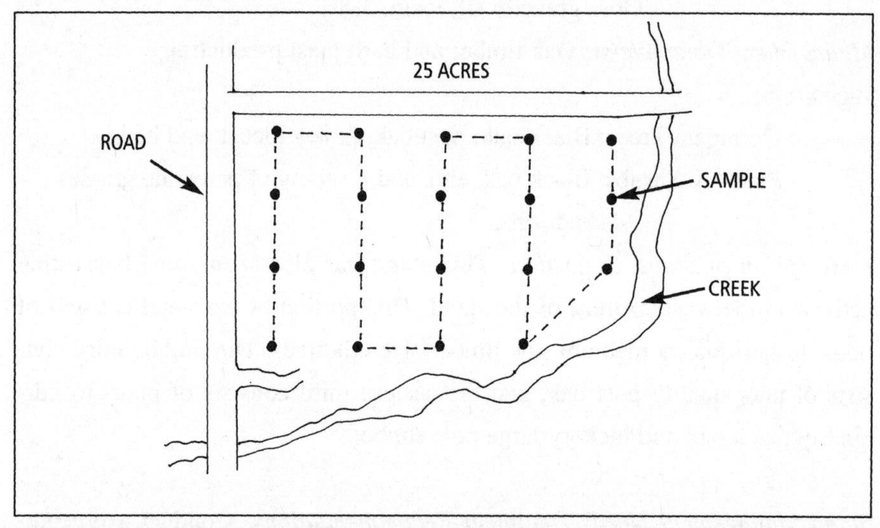

This grid pattern shows how to measure a sample of trees on a piece of property.

A cruising stick makes the chore of sizing trees easy. It can be used to measure the diameter of standing trees at breast height (DBH).

A cruising stick can also measure the number of logs in a tree.

grid can be complicated, and you'll probably want to have a professional forester do it for you.

Learn to Use a Scale Stick

To perform a timber cruise, you'll need a scale or cruising stick. These sticks resemble yardsticks but are marked with scales for measuring trees and logs. They usually have three scales: one that measures tree diameter in inches at breast height (DBH) (the Biltmore scale); one for determining the number of logs in a standing tree; and one for determining the board foot volume of a log.

Measuring diameter (DBH): To measure a tree's diameter, hold the stick against the tree horizontally about 4½ feet from the ground and about twenty-five inches from your eyes (at about arm's length). Look at the center of the tree, and then without turning your head, move your eyes to the left and move the stick so the zero lines up with the left side of the tree. Then, still without moving your head, shift your eyes to the right and read the number where you see the right edge of the tree. This is the DBH in inches.

Determining the number of logs in a tree: Pace off the base distance indicated on the stick. Find the spot at the top of the tree where the trunk becomes too small to be merchantable—four inches for pulpwood and eight inches for sawlogs. Hold the stick vertically about twenty-five inches from your eyes (at about arm's length) and adjust it so the zero at stump height, usually about one foot above ground level. Then, without moving your head, shift your line of

These days, managing for wildlife can often mean profits from the woodlands. For instance, firewood can be utilized from timber thinning.

sight to the spot you've determined as the merchantable height. The number at the merchantable height level indicates the number of logs in the tree. If it's more than one log, read to the next higher half log.

Determining the board feet in a tree: By using the DBH, the number of logs in the tree, and the board-foot scale on the stick, you can estimate the board feet of volume in a tree.

Wildlife and Profit

In the past, timber management has been geared primarily toward making a profit from the land, whether the timber is hardwoods or pine plantations. In recent years, however, interest in forest management for wildlife has grown. Forests managed for wildlife can also provide financial returns, but whether they do depends on whether the landowner has a commercial hunting or similar operation or is managing the land for personal use. Regardless of whether it's

Many timber landowners have discovered other venues, such as using a small portable bandsaw mill to mill lumber for home projects or to sell.

used commercially or privately, land that is managed for wildlife has higher resale value these days than land that isn't. Many management methods used primarily to increase timber value, such as thinning, fertilizing, tree maintenance, and tree planting, also benefit wildlife.

For example, in a thinning program, the trees selected for removal are often of little value as sawlogs, but they can often be used for firewood. This is an excellent example of earning timber profit and benefitting wildlife, because if done correctly, the operation can increase both wildlife capacity and timber value. Carried out improperly, however, the operation can do more harm than good.

With an assessment and management plan, you're ready to begin proper management of your woodlot or timberland. The following chapters describe management techniques.

CHAPTER 10

Timber Stand Improvement for Wildlife

The main goal of timber stand improvement (TSI) is removing selected trees from a stand of timber to improve the health and growth rate of the remaining, more desirable, trees. Unmanaged timber often becomes overcrowded. There may not be enough water, nutrients, and sunlight to provide optimum growing conditions for the number of trees. TSI reduces the competition in a stand and allows better growth, more mast on mast-producing trees, and more undergrowth of browse—in much the same way as weeding helps your garden. All of these improvements provide better habitat for deer and in most instances also for turkeys and other wildlife. Which trees are selected depends on several factors, including the overall tree population and whether you wish to manage for timber harvest, wildlife, or a combination of the two.

As a rule, cutting trees permits growth of more wildlife foods because more sunlight reaches the forest floor. Cutting too heavily, however, may be followed by dense sprout growth, which in general makes a poor habitat for game except in early succession or while the sprouts are young. Overcutting or clearcutting a handful of small areas scattered throughout the timber, however, can result in thickets that produce prime whitetail bedding areas as well as prime grouse and moose habitat. This should not be overdone, however. The ideal is frequent light logging cuts to maintain a stand of an uneven age and of several kinds of trees.

TSI or timber stand improvement is a common practice in timber management. Selected trees are removed from the stand to improve the health and growth rate and in the case of mast trees, bigger crops.

One of the best plans for TSI is to remove trees around better-quality oaks so they can produce more mast. Cull trees that shade acorn- or fruit-bearing trees. Because the acorns from white oaks mature in one year and those from red or black oaks mature in two years, it's important to leave many of all three varieties to maintain a yearly mast production. Dogwoods, persimmons, cherries, mulberries, hackberries, blackgums, and other trees also produce fruits that attract wildlife, including squirrels and turkeys, and elms and hickories are important to squirrels and other wildlife. If you manage for timber sales only, you'll eliminate many of these trees, but if you manage for wildlife, some of them should remain. Cut them only as suggested by your forester and according to a wildlife management plan. Remember, TSI should be done only in conjunction with a timber cruise and advice from a professional forester who will identify and mark trees to cull.

Timber is essentially an agricultural crop, and like other agricultural crops, it requires care—it needs to be cultivated. When a stand of timber is planted or when it sprouts after a clear-cut harvest, there may be as many as five thousand seedlings per acre. By the time the stand is mature and ready for harvest, each acre should have less than one hundred trees. The cycles of nature will thin surplus trees, but this process may take nearly two hundred years if unaided.

Scheduled thinning—cutting trees from an immature stand—speeds up the maturing process and improves tree quality. With hardwoods, thinning may reduce the maturing time to less than a hundred years. Thinning increases the growth rate and improves the form of the remaining trees. Proper space between trees varies according to the species, purpose of management, and quality of the site. In general, the tallest, most desirable trees are left.

The ideal stand to thin consists of trees from four to ten inches in diameter at breast height (DBH). Trees of this size respond rapidly to thinning. The thinned trees also provide firewood in ideal sizes. Stands averaging twelve inches or more can be thinned, but they don't respond quite as quickly, and many of the trees may be approaching sawlog size. If this is the case, thin only the inferior-grade trees.

Another TSI practice is *release.* Older, undesirable overtopping trees are removed or killed to encourage faster growth and better quality for the more vigorous young desirable trees in the immediate area.

Tree Selection

Although you will probably rely on a professional forester to choose which trees to cull and which to leave, you should be aware of several general rules of TSI. Walk through the woodlot and look carefully at the form, condition, size, and spacing of the trees. If you aren't good at identifying trees, carry along a field identification book. A yardstick can be used to measure trees, although a professional cruising stick is better suited for the purpose. Everything may look the same at first, but you'll soon begin to recognize patterns.

Place the trees in three categories: *final crop trees, thinning trees* that can be removed in successive or future thinnings, and *surplus or cull trees* that should be removed in the first thinning (also called a release–any tree killing or cutting to thin a woodlot).

Final Crop Trees

Trees selected for final cropping or sawlogs should be of a desirable species. They should be tall, straight, with clear trunks—no insect damage, disease, fire scars, decay, or mechanical wounds. The crowns should be full and healthy with no large dead branches, and they should be at a height level with or above existing crowns. (Those with lower crown heights due to competition may not respond as readily to thinning.) It is also important to remember that certain species do best on certain sites. In hardwood thinning, the most valuable species includes black walnuts, white oaks, black oaks, red oaks, ashes, silver maples, yellow poplars, and sweet gums. Species usually considered inferior include honey locusts, blackjack oaks, Osage oranges, and American elms. In pine plantations, species selection is usually not an issue.

Trees selected for removal may be saw logs, thinning trees, or cull trees.

FUTURE THINNING TREES

It is generally not desirable to overthin on the first cutting, as this reduces the production of wood that can be used for firewood in the future. In an overthinned stand, the quality of the remaining trees may also decline due to wind damage, persistent lower-limb growth, and less height growth.

CULL TREES

These are the trees removed in the first thinning or release operation as well as those removed in subsequent thinning. In either thinning or release methods, several types of trees are removed:

- Cull trees or widely spreading "wolf" trees.
- Trees of inferior species.
- Trees that interfere with the growth and development of selected desirable trees.
- Trees that are diseased or serve as breeding grounds for undesirable insects. Swellings or bumps on trees can indicate internal damage.
- Trees damaged by fire, scarred, broken off, bent, or seriously barked.
- Trees that have multiple sprouts from one trunk. Multiple-stem trees are quite common in many hardwood timber stands, and sometimes they can be

salvaged. Multiple stems can result from logging or from fire, and multiple-stem trees can be developed into quality single-stem trees if they're treated early enough. Multiple-stem trees are best treated when they are three inches or less DBH and less than twenty years old. At that size and age, the wounds heal quickly and it's fairly easy to discern the best sprout. If several sprouts originate from a large stump, select a low sprout and cut off all the others. When multiple sprouts are joined tightly at the base, it's harder to remove some sprouts without killing them all. In this case it's best to develop a nearby single-sprout tree. If the sprouts are joined in a low U-shaped crotch (large enough to place your foot in), remove all but the best sprout.

- Trees that are low-forked or crooked. Double-stem trees less than twelve inches in diameter with V-shaped crotches should be culled. Larger ones should be left as sawlogs. Treat sprouts on trees of inferior species with an herbicide to prevent resprouting.

Don't cull the "brush"—often small saplings or brush, are the next generation of trees waiting for the dominant trees to die, thus providing them with the sunlight and room they need to grow.

Tree Spacing

Determining spacing is done by a couple of methods. The most obvious and easy method is to determine the competition for growing space. As a general rule there should be five to eight feet of open space on at least two sides of the crowns of desired trees. This provides enough sunlight for good growth rates.

The *diameter-times-two* rule can be used in stands of trees with fairly uniform diameters. Trees are measured at breast height, and the average diameter in inches is multiplied by two. That number of feet is the ideal spacing between trees. For instance, an average diameter of six inches would call for spacing of twelve feet.

Both of these rules are general and can be broken. Trees do not grow evenly spaced unless they've been planted that way, and you may decide to leave two good trees with crowns touching if you remove other trees from around them to provide growing room. On the other hand, in extremely crowded stands it may be necessary to remove cull trees plus a few desirable trees to achieve proper spacing. The best you can hope for is average spacing.

These rules are used only to control the crowns of the main canopy trees. The understory trees are already deprived of sunlight, and culling them won't have much

The main goal in most instances is to open the tree canopy. One rule of spacing is diameter-times-two. The trees are measured at breast height. A tree with a six-inch diameter would call for twelve-foot spacing.

of an effect on the stand. Understory trees that are large enough for firewood can be cut for that purpose, however, unless this creates over-thinning.

Multiple stem trees are common after fire or logging. They should be treated before they reach three-inches DBH.

Pine Plantations

Traditionally, the pine plantations of the South have been planted for maximum timber production, and management has involved planting high densities, doing only light thinning, and no fertilizing. With this type of practice it usually takes at least thirty-five years for trees to be ready to harvest. Many landowners who desire more wildlife have discovered that newer management practices designed primarily for wildlife improvement also improve timber production. These progressive timberland management practices involve manipulating the crop to create the best final-harvest trees. They include planting seedlings with good genetics, thinning aggressively, fertilizing frequently, and pruning properly. These practices produce high-quality, harvestable timber in just over twenty years and in the process produce more open spaces for wildlife. The openings created between the trees bring in more sunlight and allow grasses, forbs, honeysuckle, and other forages to grow. To keep back hardwood sprouts and unwanted plants, landowners use prescribed burns

A V-shaped crotch or double stem tree of small diameter should be culled.

and/or herbicides. One herbicide used is BASF Arsenal, which kills the low-value plants but allows growth of the valuable forage species.

Trees to Leave for Wildlife

In commercial thinning operations many trees desirable to wildlife are removed. If you use TSI for both commercial purposes and wildlife, or primarily for wildlife, some of these trees should remain. Tall fruiting trees such as mulberries, hackberries, black cherries, black gums, and persimmons should be left. Understory flowering trees such as dogwoods, redbuds, serviceberries, hawthorns, and black haws should also be left. These trees do not compete with the main stand for light, and their fruits are valuable for wildlife.

"Wolf" Trees

Wolf trees are large, spreading trees that dominate the canopy. They usually have short but large-diameter trunks of little sawlog value and a wide, spreading crown. If they are good mast producers and are located in a stand of primarily young trees, then leave them. A good rule is to leave at least one of these trees for every two acres.

"Snag" and "Den" Trees

Snag and den trees provide valuable cover and food for many species of wildlife.

Snags: Snags are dead trees that are still standing and, in most commercial thinning processes, they are removed. As a tree begins to die and decay sets in, many species of birds inhabit it, including woodpeckers that peck out holes for their nests. These nest holes are later used by other creatures including screech

In commercial thinning, many trees desirable to wildlife are removed. These trees include dogwood, mulberry, and other fruiting trees. Leave these trees in a wildlife-based management plan.

owls, chickadees, and kestrels. Squirrels, raccoons, and other mammals also use the holes as dens. In the past, snags were removed as a part of TSI because of the potential for disease and insects. In recent years, however, foresters have begun to realize the importance of snags in the overall health of a stand of trees. Birds eat the insects that grow in the snags, and this can help prevent disease and insect problems in surrounding trees.

Wolf trees are also commonly culled during commercial operations. They do, however, provide lots of wildlife food. A good rule of thumb is to leave at least one for every two acres.

How many snags should you allow? A forester suggested the following guidelines: Leave at least one snag twenty-inch or larger DBH per acre. This will be used by red-headed and pileated woodpeckers as well as numerous other birds. Leave four snags ten- to twenty-inch DBH for species such as American kestrel and flying squirrels. And leave two snags between six- and ten-inch DBH for eastern bluebirds and black-capped chickadees. Use these numbers only in fairly large forested areas, not in those that have limited forest, including along streams or fencerows or in small, isolated woodlots.

If snags don't exist, you can create them in cull trees. Kill the live trees using an axe or chainsaw to girdle or make frilling cuts. Some trees, such as locust, also require the use of a stump killer herbicide in the frilled cuts.

Den trees: Den trees are also extremely important for a wide variety of wildlife. Most woodlots don't have enough tree cavities to provide good wildlife habitat. Many wolf trees are either den trees or potential den trees. In making a timber assessment with wildlife in mind, make sure to inventory den trees. Look for signs of rot, including decaying branches, fungi, wounds and scars, and, of course, openings or cavities. A tree with a rotted or broken-off top stem is not desirable because rain can fall into the tree. Signs of woodpecker activity indicate good den trees. The long-lived species, such as oaks, hickories, cottonwoods, black gums, and ashes, are all good choices. Leave den trees in small woodlots and along stream bottoms and fencerows as well as in larger acreages of timber.

Snag and den trees are also extremely important.

Suggested minimum numbers of den trees per acre: one tree larger than twenty-inch DBH, which will often be used by barred owls, fox squirrels, and raccoons; four between ten- and twenty-inch DHB, for gray squirrels and red-breasted nuthatches; and two between six- and ten-inch DBH, for house wrens, tufted titmice, and bluebirds.

Leave den trees during regeneration or thinning cuts. If no den trees exist, you can create them by wounding selected trees, though it may take several years for a cavity to form. Trees can be wounded in several ways. One method is to bore a two-inch-wide hole at least three inches deep into the tree. Another is to cut a limb about six inches from the trunk or chop a six by six-inch section of bark from the trunk.

Snags can be created by killing a selected tree. Girdle the tree and use a stump killer herbicide in the frilled cuts.

CHAPTER 11

TIMBER HABITAT MANAGEMENT PRACTICES

In addition to the TSI (timber stand improvement) practices described in the previous chapter, a number of other practices can also improve timber stand or woodlot habitat for a variety of wildlife. They include creating natural clearings and thickets, improving forest edge, and building brush piles.

CREATING AND MAINTAINING CLEARINGS

Although creating openings in timberland for food plots was discussed previously, creating openings using natural plants and plant succession is also important. This provides a wide diversity of plant species that will attract many types of wildlife.

Large areas of timberland are essentially a monoculture—they don't provide the variety of plants that will encourage a more diverse wildlife population. "Naturalized" clearings created in timberland can provide one of the most important factors in wildlife habitat: edge. Many biologists suggest that even a clearing as small as an acre can provide as many edible plants as ten acres of timberland, and the grasses, seedlings, forbs, and

Vast areas of monoculture timberland are not as attractive to many wildlife species because they don't provide a diversity of plant life. Creating clearings improves habitat for game such as deer, turkeys, squirrels, quail, bear, moose, and ruffed grouse.

annual weeds provide nesting and cover for birds.

As a general rule, one hundred acres of timberland should have five to ten acres of clearings. Each clearing should be fairly small, from one to three acres. The ratio of timberland to clearings depends on the makeup of the timberland and the surrounding habitat. Small woodlots in agricultural country or woodlots with ungrazed fields around them do not need as much land devoted to clearings as do large blocks of woodlands.

Clearings You May Already Have

Utility or power line right-of-way. These areas must be kept open—food-plot planting isn't allowed—but some management practices not only help to keep them open but also increase wildlife usage. Check with your utility company about what practices you can follow to improve wildlife use of right-of-ways on your property.

Woodland roads and trails. You probably already have them on your land, they're probably aleady used by wildlife. We have about five miles of roads and trails through our woodlands. They are used by humans not only as access to different parts of the property but also as fire roads and fire lanes. We also have several natural openings scattered throughout the woodlands. Deer feed in these areas regularly and turkeys use them for bugging, strutting, and displaying.

One clearing that can often be enhanced is a utility right-of-way (with permission from the utility company).

Temporary clearings. These are created by practices such as logging or by natural disasters—fires, tornadoes, or hurricanes. Temporary clearings can be planted in annuals such as lespedeza and wheat until they regenerate naturally, or they can be replanted in trees.

Permanent clearings. These may be created the same way as temporary ones are, though more often they're clear-cuts. These permanent openings provide the most food for wildlife during the first five years after they are created. To maintain high-quality food sources, the clearings must be kept in early succession or with young plants and not allowed to mature to provide a continuous supply of natural legumes, annuals, perennials, and small woody vegetation. This can be done by mowing, burning, using selective herbicides, or a combination of these practices. It's a good idea to check these clearings about every three years to make sure they are providing a good mixture of foods.

Creating Clearings

If you don't have natural clearings, you'll have to create them. In addition to roads and trails, other types of openings can be created.

Types of clearings

- *Clear-cuts.* In clear-cutting, all timber is removed to create a completely open area. Right-of-ways are examples of clear-cuts. You might also wish to clear-cut a boundary between your property and your neighbor's that will also serve as a firebreak. These types of clearings can be quite small—as little as half an acre. Clearings of one to two acres are also effective.
- *Savannahs.* The best method on larger acreages, from five to fifty acres or more, is to create a "savannah," which was a common

Woodland roads and trails can often be improved for wildlife by widening and planting food plots or encouraging herbaceous growth.

A savannah is a clearing with a few select trees left standing. In the early days, savannahs were natural habitat types along the edges of prairies and timberlands.

type of habitat before the axe, saw, and plow changed much of the makeup of our timberland. These unique habitats are mostly open with only a few large and scattered fruiting trees, usually oaks. One such area on our place is alive with deer when the white oak acorns begin to fall each fall, and the gobblers strut in it each spring.

Methods of clearing

- *Removal.* You can create a clearing with a chainsaw and/or brush saw, tractor and blade, or dozer.
- *Girdling or frilling.* One of the simplest methods, although not pretty, is to kill the standing timber by frilling or girdling the larger trees. As the trees die, the openings created allow more sunlight in and this increases the production of forbs and other plants. The standing dead snags also provide food and nesting for various types of wildlife, and as the limbs fall to the ground, even more cover is added. It is difficult to keep this type of clearing in first succession, however, because you can't rotary-cut it, but you can burn it annually.
- *Herbicide.* An extremely effective method of creating a clearing is to use a selective herbicide. Arsenal Applicators Concentrate (AC) herbicide is an environmentally friendly herbicide that kills weeds and brush but is safe to humans and wildlife. Because the active ingredient in Arsenal AC works by inhibiting an enzyme

system present in plants but not found in mammals, fish, birds, or insects, it is essentially nontoxic to humans and wildlife when used in accordance with label instructions. Arsenal also does not accumulate in the tissues of mammals, birds, or fish. Arsenal AC herbicide is an aqueous solution that is mixed with water and applied to control most annual and perennial grasses, broadleaf weeds, and brush.

The following habitat enhancement projects can be achieved by using Arsenal. Whether these are applicable to you depends on the amounts and types of forest and vegetation on your property as well as your specific wildlife management plan.

- *Combatting overgrowth in clearings and old fields using Arsenal.* More than one new landowner has discovered that the back forty field described by the previous owner as a "food plot" is actually a thick stand of young low-grade hardwood sprouts. You can clear the area with a brush-hog or bulldozer, but the stumps will quickly resprout unless you treat them. Arsenal AC can be used to control this undesirable vegetation. Apply a solution of the herbicide and water to the cambium area of the freshly cut stump surfaces or to cuts on the stems of the targeted woody vegetation.
- *Killing overgrown stands of Osage orange (hedge) and locusts.* Arsenal AC can also be used on these problem trees, which can easily spread over fields and clearings. Make one hatchet cut for every three inches of diameter of the tree at breast height and inject one milliliter using standard injection equipment in each of these cuts. Targeted trees can also be frilled or girdled. Using a hatchet, machete, or similar tool, cut through the bark at approximately equal intervals around the tree. Spray or brush the Arsenal AC solution into each cut until it's thoroughly wet. A hand-pump spray on a plastic bottle works well. Incidentally, this is also an excellent way to create den trees for cavity-nesting wildlife such as squirrels.
- *Managing forest understory.* In addition to creating clearings, managing the forest understory—the forest's woody brush—can also attract more wildlife such as deer and turkeys to your land. Some woody brush is great, but too much can be a problem.

Often the understory is of low food quality, and it overshades higher-quality foods such as forbs, legumes, and other natural food plants. In the past, understory management was primarily carried out by prescribed burning, and this is still a very viable management tool. Burning is a relatively short-term solution to the problem, however, because low-quality hardwoods will often resprout quickly. Using Arsenal herbicide treatments between burns can reduce the number of burns you'll have to make to one every five to seven years instead of the usual frequency of one every two to three years. The intensity of burns is also substantially less when you use Arsenal, resulting in less likelihood of fire getting out of hand because the quantity of brush is less. Using an herbicide between burns also greatly reduces the potential for wildlife fire losses because the majority of the fuel—understory and midstory, low-quality hardwood—has been removed.

A number of university studies have proven that overall biodiversity is not only maintained but often greatly enhanced through the use of Arsenal AC. There is a shift in plant species recolonization from low-quality hardwood brush to high-quality forbs, legumes, and rubus species (dewberrry and blackberry). As much as a thirty-three-fold increase in forbs and a doubling of vines such as honeysuckle was measured in one Mississippi study.

Forbs make up a very significant portion of the diet of white-tailed deer. According to the Mississippi study, they make up as much as thirty-three percent of the vegetation deer consume, so increasing the quantity of forbs can be a major factor in a deer management program, even on a small scale.

Leguminous hardwoods, including redbud and soft, mast-producing species such as persimmon and dogwood are tolerant of Arsenal, so when the herbicide is used, they may become more abundant, attracting more wildlife. Bugging opportunities for ground birds such as quail and turkeys are also improved. (Northern bobwhite quail tracking studies indicate that quail prefer treated areas over untreated areas due to the increase in flowering plants, which greatly increases the number of insects present.) Individual plant seeds in treated areas have been found to be larger and to have a higher protein content than in untreated areas, which translates into higher-quality foods for deer, turkeys, and other wildlife.

This doesn't mean, however, that you should remove all brush. Deer need edge and brushy bedding because travel and hiding areas are very necessary for them, and turkeys and quail also need protective cover to some degree. The key is to control the amount of brushy cover and provide more food availability at the same time.

- *Southern pine Arsenal experiment.* One unusual experimental use of Arsenal in a Southern pine plantation was the creation of a "stalking" trail—a maze-like opening with lots of twists and turns. Open areas were created alongside it so that bowhunters could stalk the trail for a unique hunting experience.

The understory of the woodlands should also be managed to provide optimum habitat. Some woody brush is great, too much can be a problem, depending on the species of wildlife the area is managed for.

Maintaining Clearings

- *Mowing.* We maintain roads, trails, and clearings with a rotary mower in late summer or early fall, well before deer begin creating scrapes. In fact, we usually see scrapes within a week or so of cutting back these clearings.
- *Burning.* In the past, nature or Native Americans maintained these natural clearings by the use of fire, and a prescribed burn is still one of the most effective means of maintaining them. (See Prescribed Burns.)

Improving Forest Edge

The quality of forest edge—the land that borders woodland—is extremely important. The idea is to create a transition zone between the forest and a clearing, pastureland, agricultural cropland, or a food plot.

Without proper management, most forest-to-opening zones are quite abrupt. There are tall, closely spaced trees, then open ground. Most crops won't grow very well within twenty feet of the abrupt edge of a forest due to shading, lack of water, and root competition.

The ideal situation is to create a thirty-foot or wider transition border between the woodland edge and the open ground. This can be composed of grasses, shrubs, vines, weeds, brush, and small trees. If you're a deer hunter, you'll know that deer use these brushy areas as lookouts before they move into a field to feed in the evening. The areas also offer a variety of foods and nesting cover for birds.

Creating clearings can be done by clearcutting, bulldozing and/or using herbicides.

A transition zone can be created in the field or in the forest's edge. Creating one in a field can be as simple as allowing the area to grow back in native vegetation, or the area can be planted in shrubs, vines, and other plants. If you create a transition zone in timber, use the same techniques you would for creating timber clearings, including clear-cutting or selective cutting. All large trees within thirty feet of the outside edge must be removed or killed by girdling

One of the most important practices is managing forest edge so there is no abrupt edge between forest and grass or croplands. A thirty-foot wide transition area of brush, vines, and shrubs is ideal.

and herbicides. Some trees can also be cut low to promote sprouts. Thinning can create a gradual regrowth of forest transition edge. If you choose this plan, leave some snag and deadened trees within the edge, and when you cut trees with vines on them, don't cut the vines. Don't worry about creating a perfect, uniform-width strip. The wildlife won't care, and in fact an irregular strip is preferable.

The above techniques will bring you closer to the ideal situation—a well-managed woodland with a diversity of hard-mast and soft-mast trees of varying ages separated from an agricultural field, ungrazed grassland, or food plot by a diverse woodland edge.

Creating Brush Piles

Whenever you cut trees down, be sure to leave the treetops behind. Loose treetops provide nesting areas for turkeys and protection from predators for quail and turkey poults, and larger brush piles afford cover for rabbits and other small game as well. If you cut trees in late summer, not-quite-developed mast will be at ground level, where it will be attractive to turkeys and deer.

More detailed information on creating brush piles is covered in Chapter 29.

CHAPTER 12

REFORESTATION

Generating a new forest is a standard practice for tree farmers and can also be a great wildlife habitat practice. Reforestation can be done in an area where no forest exists or regeneration of a forest can be done after a mature forest is cut for market.

CUTTING METHODS

Four basic cutting methods are used for regeneration: selection, shelter-tree, seed-tree and clear-cutting. These methods differ primarily in the amount of sunlight that reaches the forest floor after the cutting is done.

Regenerating a forest, or creating a new one by planting, is a great way of creating wildlife habitat. Several methods may be used depending on timber harvest and wildlife management goals.

The characteristics of the land—soil, topography, moisture, and microclimate—are also important in choosing the right system, as are the capabilities of the existing vegetation, including seeding habits, growth patterns, and "tolerance"—the ability of a tree to grow in the shade of other trees.

It's also important to consider whether you desire uneven-aged or even-aged timber. Even-aged forests are common in commercial timber operations, particularly in pine plantations. The best forest system for wildlife management, however, is an uneven-aged system. It's important to choose the right method for your particular forest and the goals you wish to attain. A professional forester is invaluable for accessing and providing information.

Selection Method

Individual trees or small groups of trees are selected for cutting. These cuttings may be made each year, but are normally made in intervals of five years or more. The large, mature trees are harvested for wood products, and this creates openings for new growth. The smaller cull trees are also removed, including diseased and poorly formed trees and those of undesirable species. If you're managing for wildlife, you may wish to leave some cull-type trees, but you should remove enough trees to reduce crowding and provide sunlight to the forest floor to stimulate the remaining trees. Small groups of trees, up to one or two acres, may be cut rather than just individual trees. This creates slightly larger openings in the forest and also creates uneven-aged stands. Using the selection method, the forest is continually regenerated by new trees that spring up or sprout where the mature trees were removed. Shade-tolerant species are best suited for this system.

Shelter-Tree Method

A new stand of trees is established under the shelter of old trees. This method is best suited for tree species that do not compete well with other vegetation in direct sunlight, especially in the early years. Two to three cuts are commonly made. The first cut stimulates seed production in the trees that are left, prepares the ground for the seed by stirring up the soil, and lets in some light. In this type of cut enough older trees are left to shade the site for at least part of the day. Sprouts may grow from cut stumps, natural seedlings may sprout, or you may plant seedlings. A second cut (anywhere from three to ten years later) is then made, and sometimes a third, to remove the remaining mature trees, completely releasing a young stand. Once the mature trees are cut, an even-aged stand remains. One of the disadvantages of this method is that some young trees may be damaged in the second and third cuts.

Seed-Tree Method

The majority of the mature trees are removed in one cut, leaving only a few mature "seed trees" scattered over the area. Once the young trees have established themselves, the remaining mature trees are cut. This method should be used only with firmly rooted species that will not blow over when the forest is opened up and exposed to strong winds. Intolerant species such as oak do best in this system. The final result in ten to twenty years, depending upon the species, is an even-aged forest.

Clear-cut Method

A clear-cut is a regeneration cutting. It is normally done on timberland with enough marketable trees to produce a financial gain. Clear-cuts may be of two types: commercial or silvicultural.

Commercial. Only the marketable trees are removed, leaving the unmerchantable trees standing. They are often inferior, suppressed and/or shade-tolerant species that can create problems in regeneration, especially if the purpose of regeneration is to attract wildlife. In some instances natural regeneration is used on this type of clear-cut—the seeds from the merchantable trees are allowed to regenerate the forest. It takes a fairly long time for this to occur naturally .

Silvicultural. (Silviculture is the science, art and practice of caring for forests with respect to human objectives.) All trees larger than one or two inches in diameter—both merchantable and unmerchantable—are removed. The unmerchantable trees are used for chips. Silvicultural clear-cutting may be more expensive initially but offers several advantages for wildlife habitat creation. First, there is no competition for seeds or seedlings, resulting in faster growth. Second, the slash left after the cutting gradually breaks down in the soil, offering nutrients to the new seedlings. This type of clear-cutting produces even-aged stands. Although uneven-aged forests are generally better for wildlife, some clearcutting is a good practice for some species, such as deer and bear, and is necessary for other species, including grouse and moose.

Regeneration after a clear-cut. Regeneration after clear-cutting can be done several ways depending on the cutting system, your goals, the amount of money you have to spend, and the amount labor you're willing to devote. The methods include *natural ("coppice") regeneration* (you allow seeds to sprout and sprouts to grow from cut stumps of hardwoods) and *artificial regeneration* (you seed directly or plant seedlings).

Natural regeneration. Natural regeneration is easier than artificial regeneration, as you simply allow the site to regrow, but even in a natural system you'll need to do some work. Remove all but one sprout from each cut stump. You may also need to fertilize, prune, and thin undesirable species like shade-tolerant hickories, which may grow in stands so thick a rabbit can't wiggle through.

Artificial regeneration. Artificial regeneration provides quicker reforestation and in the long run offers more return on your money than natural regeneration does. The key to artificial regeneration is using genetically improved seeds or seedlings. Site preparation, the cost of seeds or seedlings, and the amount of labor for planting, of course, are greater than with natural regeneration, but with more control over the plant species you can regenerate a forest designed for a specific wildlife habitat and at the same time grow financially marketable timber. Normally the trees are planted eight to ten inches apart at the same depth or slightly deeper than they grew in the nursery.

- *Direct seeding:* In natural regeneration, thousands of seeds are broadcast, but only a hardy few survive. In the past artificial regeneration was accomplished by planting seedlings. As many as a thousand seedlings were planted in a single acre with the hope of harvesting approximately one hundred of them. In recent years a method called direct seeding, which is similar to natural regeneration, has evolved. In this method germinated seeds or "nutlings" are planted. Direct seeding offers several advantages over both natural regeneration and planting seedlings: selected species can be planted (for example, you may wish to plant sawtooth oak, a tree widely used by both white-tailed deer and turkeys); it's easier than planting seedlings; and it's more cost-effective than planting seedlings.

  To succeed in directing seeding, follow these suggestions: plant genetically superior "nutlings"; use a slow-release fertilizer; most importantly, use a protection system to allow the nutlings to grow into trees. Doing all of these is extremely simple. Place the nutling inside a specially designed tree protector; add a bit of mycorrhizal inoculant (sold with the nutlings); then add a bit of soil. Direct seeding products are available from Quality Deer Management Association. They include species such as sawtooth oaks, gobbler oaks, white oaks, persimmons, crabapples, black walnuts, chinquapins, and hawk chestnuts. Packages come with tree protection systems and slow-release tree fertilizers.

- *Planting seedlings:* The most common method of artificial regeneration is planting seedlings. Seedlings are available from a wide variety of sources. Many states offer tree seedlings of species that are especially useful for wildlife habitat enhancement. The Quality Whitetail Deer Management Association and the National Wild Turkey Federation (NWTF) offer seedlings, as do numerous private forestry firms and nurseries.

Before you plant, it's important to determine which types of trees will grow well in your area or on your specific site and what species are best suited for your management goal. A wide variety of tree species can be used for creating a wildlife forest, including sawtooth oak, northern red oak, white oak, pecan, black walnut, pin oak, white mulberry, white pine, and Chinese chestnut. If you wish to attract deer, you may also wish to add apple trees. Stark Bro's has a wildlife special of ten apple trees of two hardy, Midwest-grown varieties. The bare root trees are five to six feet tall when shipped and will grow to a mature height of twelve to fifteen feet. Find out what species grow well on the specific soil type and topography of the site you're planning to reforest. As you walk the site, note the soil type, present ground cover, and soil drainage. (Most trees do poorly in soils with poor drainage, but some do well.) Local foresters and university extension experts can help determine the best species for your area.

Once you know what kinds of seedlings you need, determine how many trees you can plant and properly care for. Measure the area to be planted and determine the spacing and number of trees. In an eight-hour day, two people can plant about eight hundred trees in open, easily worked land. If you have several acres to plant, you may wish to rent a tree-planting machine. An experienced three-person crew with such a machine can plant as many as a thousand trees per hour. Consider how much money you have to spend; also consider that regular maintenance will be required each year on the site.

Once seedlings arrive, it's extremely important to keep the roots well protected and moist until they can be planted. It's best to plant them immediately, but if you can't, you can store them for a week or two by placing them in a cool, shaded area. Separate the bundles to prevent overheating and growth of mold and regularly pour cold water into the open end of each bundle to keep the roots moist. Don't submerge the roots in water, however, as they will become damaged. Protect the seedlings from freezing.

If you need to store the seedlings for more than a couple of weeks, keep the bundles in cold storage at 35 to 37 degrees F, but do not allow them to freeze. You can also place the bundles in a heeling trench. Dig a trench deep enough to cover the roots in a shaded and protected place. Don't do this in areas with high rodent populations, as the rodents will chew the roots and bark from the seedlings. Cut the strings holding the bundles and spread the seedlings out in the trench. Water as needed to maintain soil moisture around the roots, but do not overwater. Placing mulch around the seedlings will help maintain moisture and add further protection. If the trees begin to grow before you plant them, leave them in the heeling bed until the next spring. Do not allow open bundles of seedlings to be exposed to the sun and wind, even for short periods of time.

When planting seedlings, take only a few bundles of trees at a time. Leave the others covered and moist until you're ready to plant them. Carry seedlings to the planting site in a bucket half-filled with water or in wet moss. Be careful as you plant to avoid damaging the terminal buds on the seedlings.

-Hand-planted

Hand-planted seedlings may be placed using the center or side-hole method or the slit method. Whichever method you use, on unprepared sites you must first scrape or "scalp" the sod, grass and weeds in an eighteen-inch-diameter circle.

With the center or side-hole method, use a hoe or mattock to make a hole in the center of the eighteen-inch circle. This is the best tool and method for rocky soil. Drive the blade of the tool full-length into the soil with the handle parallel to the

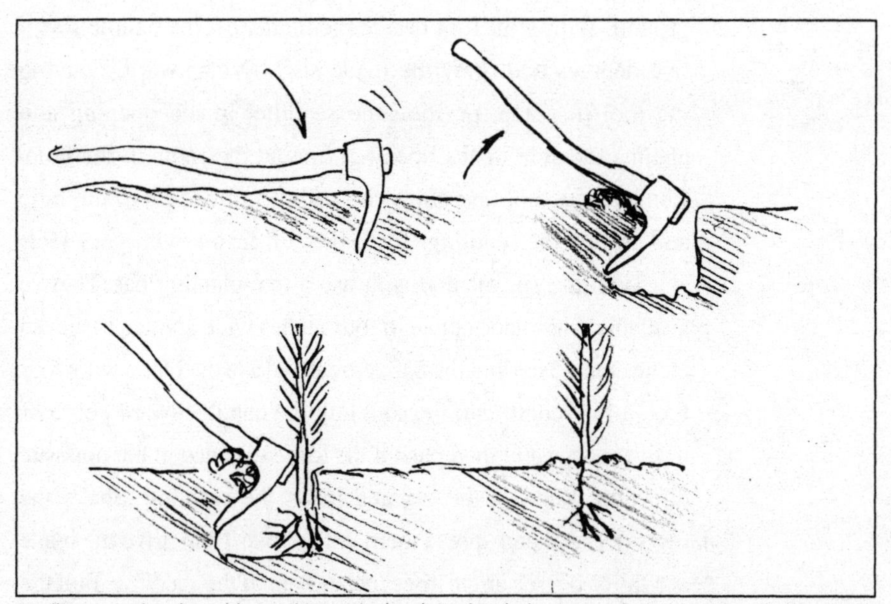

Seedlings may be planted by machine or by hand. For hand planting, scalp an area in an eighteen-inch circle where the tree will be planted. A mattock can be used to create a hole. This is a good choice, especially for rocky soil.

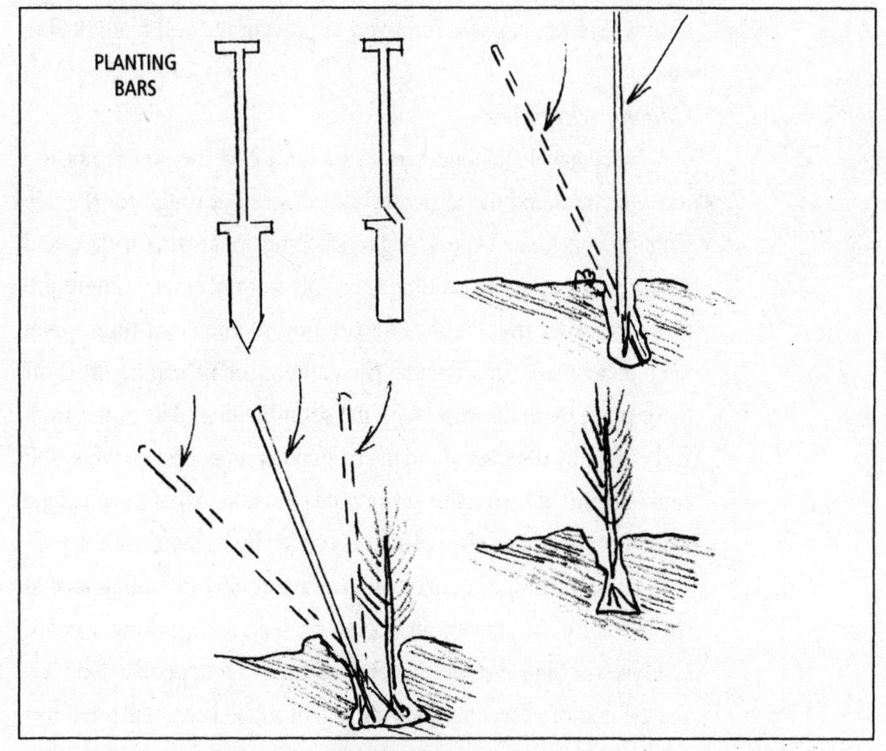

A tree planting bar is used for the slit method. The bar is driven into the ground, pushed back and forth to open the hole, the seedling placed in position, then the bar used to push soil against the seedling.

ground. With your foot beside the blade, lift the handle forty-five degrees and draw the blade slightly backward. Leaving the tool in place, position the seedling in the opening and against the side of the opening. Spread the roots in the hole. Remove the tool and allow the soil to fall back into the hole and around the seedling. Tamp the soil firmly with your feet.

With the slit method, you use a tree-planting bar. They're available with blades three to four inches wide and ten to eleven inches long. Holding the bar vertically, drive the blade with your foot, a full length into the soil. Pull the handle toward you four to five inches and then push it the same distance in the opposite direction. Remove the bar and insert the seedling. Shake the roots out well and spread them in the hole. Then drive the blade at a thirty-degree angle three inches behind the seedling. Pull the bar toward you and then push it forward to close both the top and bottom of the slit. Remove the bar and use your foot to tamp the hole where the bar was removed to prevent the soil from drying out.

-Machine Planting

Mechanical planting can speed the job if the soil is not too rocky or hard and the terrain is not too steep or rough for the tree planting machine. Unless you're planting more than a thousand seedlings, mechanical seeding is usually not necessary. There are three types of machines. The floating planter is a three-point hitch attachment for a tractor. The entire machine can be lifted off the ground by the tractor. With the semi-floating planter, the front end is held by the tractor and the rear end is supported on wheels. It cannot be lifted by the tractor for transportation. All of the machine weight of the trailer-style planter is carried by its own wheels.

These machines consist of a rolling coulter (a cutting tool to slit the soil), a trencher, an operator's seat, and packing wheels. Using a machine requires the labor of three to four people. One drives the tractor; a second rides the machine and keeps the seedlings protected, separated, and loaded into the planting machine trays (the trees in the planting trays are kept covered with wet burlap or moss. If the roots become exposed to the sun and wind, the trees may die before they are planted); a third follows the machine to straighten and pack poorly planted trees. With a four-person crew one person transports the trees from storage to the planter.

Caring For New Plantations

- Keep livestock away from newly planted areas.
- Plow or disk a firebreak around the area to protect the young seedlings from fire.
- Prevent weeds and grasses from growing rank around the trees. Use selective herbicides, disking, mowing, or hoeing to keep down competition.
- Inspect for evidence of disease and insects and treat with insecticide if necessary.
- Fertilize with slow-release tree fertilizer according to the manufacturer's instructions.
- Use tree shelters to protect trees from damage by deer and other wildlife.

These translucent plastic "mini greenhouses" cause the seedling to grow faster, extend the growing season in the fall, and also make it easier to locate smaller seedlings for fertilization. Shelters are available in several sizes ranging from eight to seventy-two inches high. They should extend twenty-four inches above the top of the seedlings to be planted. They're extremely easy to assemble and use. Place the ties that come with the shelters through the pre-punched holes and fasten them loosely to bend the shelters into tubes. Drive a stake into the ground next to each seedling. Slide a shelter over the stake and seedling and at the same time slide the ties over the stake. Tap the top of the shelter to drive it about an inch into the ground. Tighten the ties around the shelter. To prevent birds from nesting inside the shelter, put a piece of netting over the top. Shelters are available from the NWTF through HELP, Habitat Enhancement Land Program.

CHAPTER 13

FOREST MAINTENANCE

Just as your yard needs constant maintenance, timberlands, to be productive both financially as well as in regard to wildlife, also need maintenance. We've already discussed thinning, cutting, and planting in previous chapters, and these are all maintenance practices. Other maintenance practices include protecting, fertilizing, weed control, and pruning.

PROTECTING

One of the most common practices among landowners, and one of the worst for woodlands, is allowing livestock to graze the area. Grazing can damage a forest extremely quickly, and extended grazing can cause damage that may last for many years. Our property was heavily grazed before we purchased it, and in some places the soil is so thin nothing but mosses will grow. Grazing woodlands causes several problems, including broken limbs, compacted and

One of the worst problems with timberlands is allowing unlimited grazing by livestock. Grazing creates erosion problems, removes vegetation and leaf cover, and can change timberland species to the detriment of wildlife.

eroding soil, and the destruction of seedlings. Unfortunately, woodland grazing seems like a good idea—there's shade, protection from winter wind, and shelter for calving. The cost of fencing woodlands may be a factor in deciding whether to allow livestock grazing. If you want to attract wildlife, however, it's a cost you must live with. Ungrazed, healthy woodlands act as giant sponges. Leaf litter and vegetation lessen the impact of raindrops: with their energy dissipated, they gently soak into the soil, preventing erosion and sedimentation. Grazing removes vegetation and leaf cover, which can increase erosion enormously. As the falling rain hits bare, compacted earth, it begins a downhill run that carries large amounts of soil with it.

Ungrazed woodlands also usually contain a diversity of tree species of various ages and size classes. This is especially true if timber stand improvement (TSI) is used to manage the timber for wildlife. Grazing can change the makeup of the woodland. The oaks, desired for their mast crops, often die as numerous minute feeder roots near the soil's surface are trampled. The less-valuable hickories aren't as affected by soil compaction, and their numbers increase. Thorny locusts also soon appear and, because of their thorns, cattle don't eat the branches, but they do eat the seed pods and further spread this species with their manure.

Grazing doesn't allow seedlings to develop to replace older trees as they die or are harvested—in fact, the young, tender seedlings are the first to be grazed. And livestock can damage the bark and limbs of healthy trees, eventually causing them to die.

Studies have shown that it takes twenty to forty acres of woodland to produce the same quantity of forage supplied by one acre of grassland. Wildlife need healthy woodlands. The brushy undercover provides food and protection for many species. Grazed woodlands become park-like with little to offer wildlife.

Fencing woodlands these days is much more cost-effective than it used to be. An electric fencing system with high tensile wire is fairly economical and easy to install. A single strand of wire is positioned about eighteen inches off the ground. This keeps the cattle in, but allows wildlife the freedom to move in and out. We completely fenced the riparian woodland border of a creek with this method.

FERTILIZING

Trees, like all vegetation, require fertilization for proper growth. To get the most from your hard-mast and soft-mast trees, set up a fertilization program. Fertilizer can be used when the tree is planted, but don't use a high-nitrogen fertilizer at that time. A good choice for seedlings is Nutri-pak tree fertilizer, available from the National Wild Turkey Federation. This slow-release product lasts one to two years and won't burn seedlings. After digging the planting hole, drop in one Nutri-pak fertilizer tablet. If you're planting balled trees, Scotts Tree Tablets are an excellent choice. Use four tablets per inch of tree trunk diameter (measured one foot above soil surface). Place the tablets evenly around the root ball and about an inch from it, about halfway from the bottom of the planting hole.

Trees, like most plants, will quite often do better with proper fertilization. Granular fertilizers may be used. Scotts Tree Tablets with timed-release formula are easy to use.

Granular fertilizer for trees, like that for other plants, is made up of three elements: nitrogen (N), phosphorus (P), and potassium (K). The percentage of each is represented by a number on the fertilizer package. For example, 12-12-12 fertilizer has twelve percent of each element and 0-15-15 has no nitrogen and fifteen percent phosphorus and potassium. Understanding how the various elements affect plant growth is important. Phosphorus and potassium promote growth in plant parts that are underground—the roots. Nitrogen encourages growth in parts that are aboveground—the trunk, limbs, and leaves. In the spring apply a general-purpose fertilizer with nitrogen, such as 12-12-12, and in the fall use one that has no nitrogen but is high in phosphorus and potassium. This promotes root growth and mast production. To help oaks to grow more mast, a fertilizer high in phosphorus and potassium can also be applied in the winter months. Nitrogen fertilizer should, however, never be applied to trees in the fall, as it prevents sap from falling. During spring, when the sap rises, the tree forces nitrogen up to promote top growth.

It's important to apply the correct amount of fertilizer. Give trees and shrubs from two to three years old and/or those with a trunk diameter less than 1 inch a couple of tablespoons of fertilizer applied around and under the drip line (the area under the canopy). For older trees and those with larger trunks, a general rule is to apply one pound of fertilizer per inch of diameter, around the inside of the drip line, but not allowing the fertilizer to touch the bark.

An easy way to fertilize trees is with Scotts Tree Tablets. The timed-release nitrogen formula lasts up to two years. The formula is 20-10-5 plus some minor elements. When you use timed-release fertilizer, trees receive a slow, gradual feeding, and there is no nutrient loss from leaching as there is with quick-release fertilizers. The nutrients are meted out gradually, as the trees need them, so they produce higher yields.

To feed established trees with Scotts Tree Tablets, use an auger or dibble (heavy bar) to make six-inch-deep, 1½-inch-wide holes every three feet around the tree, directly below the spread of the tree's branches. Place one tablet in each hole and cover with soil.

Controlling Weeds

Young trees need protection from the competition of weeds and fast-growing brush. An area of at least two feet around the tree should be kept clear of vegetation until the seedling becomes large enough to create a shaded drip line. Hand pulling, mowing, and hoeing are options on smaller plots. Just make sure not to damage the seedling. It's especially important when hoeing that you don't damage the shallow roots. Selective herbicides may also be used under low pressure. A plastic bucket can be used to cover the tree while you're spraying the herbicide.

Pruning

Properly done, pruning forest trees can yield benefits both financially with higher-grade trees and for wildlife with an increase in foods. Naturally, in a forest you'll probably prune only the higher-value trees, such as white oaks, walnuts, chestnuts, and pecans.

Proper pruning begins early in the life of the tree. If you're growing a pine plantation or regenerating a hardwood forest, pruning early is extremely important. Pruning actually wounds a tree, and it's important to understand the basics of proper pruning. Some pruning tips:

When to prune. The best pruning time is during the dormant season or in late winter or early spring before leaves form. Wounds created by pruning heal most quickly at that time, and sprouts from dormant buds are less likely to develop. Don't prune during leaf formation or during leaf shedding. Remove any diseased or dead branches any time you notice them.

Judicious pruning may be in order to have prime trees.

How to prune newly planted trees. At the time of planting, prune the tree to select one central stem. Removing multiple leaders is especially important on valuable trees such as walnuts, where you desire a single straight trunk. Once you select a strong central leader, don't prune this stem or any twigs along its upper branches. It's important to maintain at least two-thirds of the total tree height in living branches.

How to prune mature trees:

- *Prune living branches as close as possible to the trunk.* Do not, however, cut behind the branch's bark ridge. (If you look carefully, you'll notice a thick bark ridge separating the branch from the main stem.)
- *Don't create flush cuts or cuts right up to the trunk;* they generate large semicircular or horseshoe-shaped calluses. If you prune correctly, a circular callus will form around the wound.
- *Don't leave branch stubs.* They will eventually decay and fall off.
- *When removing dead branches, don't cut into the collar at the base of the dead branch.* This raised ring of protective tissue encircles the branch and is a barrier to prevent further decay.
- *Prune smaller limbs as soon as possible,* because the wounds heal more quickly than those from pruning larger branches. If you do prune large branches, use a three-step method to prevent the weight of the falling branch from tearing the bark from the tree below the cut. First make an undercut out from the branch fork.

Then make an overcut to meet the undercut. This will allow the limb to slowly fall over and release, then cut off the stub. On large limbs be careful that the limb doesn't jump back at you when the branch hits the ground.

- *Don't top trees.* This old-fashioned practice is still used, but it does trees no good. Storm-damaged upper limbs, however, should be removed as quickly as possible after injury. Cut them at about a forty-five-degree angle along the branch bark ridge.
- *Leave wounds bare.* Studies show that commercial wound dressings, asphalt, tar, and even beeswax do not help the tree to heal.
- *Use proper tools for pruning,* like pruning shears, and make sure you keep them sharp and clean. To prevent spreading diseases from tree to tree, sterilize pruning tools in a mixture of one part household bleach to ten parts water.

Threats to Woodlands

Disease

Many diseases affect woodland trees. Many of them are extremely hard to control in larger plots, although they can be controlled to some degree in smaller ones. In smaller tracts, or with particularly valuable trees, you may wish to treat diseases with fungicides. Many diseases are spread through host plants, and eliminating these host plants is one way to control diseases. Quickly pruning affected limbs from valuable trees is another.

Insects

Many insects prey on timberlands. For the most part, in healthy woodlands managed for wildlife, birds and other animals will keep insects under control. With especially valuable trees, you may choose to use an insecticide.

Parasites and Invasive Species

Some vines are parasites of their host species—for example, grapevines love black walnuts. Although the grapevines create food and cover for many wildlife species, an overabundance of them can reduce mast production. You may wish to maintain some tangles of grapevines but release other trees from their stranglehold. Do this by cutting the vine about a foot from the ground and treating the cut with stump-killer herbicide. Another problem vine is poison ivy. This not only can strangle small trees, but it also spreads rapidly. Although the seeds are a food for some birds, the vines can quickly become a nuisance, not to mention a health hazard to those who wish to enjoy the woods.

A number of other—often "exotic" or introduced—species can reduce the value of woodland to wildlife and a source of income. Some exotics can even overgrow a mature forest. They include vines, grasses and shrubs.

Vines to watch for and control include kudzu *(Pueraria lobata)*; Japanese honeysuckle *(Lonicera japonica)* (not to be confused with native honeysuckle); Chinese wisteria *(Wisteria sinensis)*; and Japanese climbing fern *(Lygodium japonicum)*.

Grasses that can become a problem include Johnsongrass *(Sorghum halepense)*; Bahia grass *(Paspalum notatum)*; giant fescue *(Festuca arundinacea)*; Bermuda grass *(Cynodon dactylon)*; Japanese grass or stiltgrass (*Microstegium vimineum)*; and cogon grass *(Imperata cylindrica)*.

Shrubs to watch for and control include multiflora rose *(Rosa multiflora)*; bicolor lespedeza *(Lespedeza bicolor)*; serecia lespedeza *(Lespedeza cuneata)*; Chinese privet *(Ligustrum sinense)* and Japanese privet *(Ligustrum japonicum)*.

Several exotic trees can also spread rapidly and include popcorn tree or green tallow *(Sapium sebiferum)*; Chinaberry *(Melia azedarach)* and silk tree or mimosa *(Albizia julibrissin)*.

Herbicides can be used to control these exotic nuisance invaders, although some plants are extremely hard to kill even with correct herbicide usage and may require several treatments. It's important to match the correct herbicide and usage to the plant to be controlled. Local county extension offices, conservation departments, and Soil Conservation Service offices have detailed information on herbicide selection and application. Additional information is also available online from university extension services and herbicide companies.

Maintenance of your woodland is part of the joy of creating better wildlife habitat. It's also a great excuse to spend another day in the woods.

CHAPTER 14

TIMBER AND WOODLOT MANAGEMENT TOOLS

One of the most important tools for timberland management is the chainsaw. Whether you're doing timber stand improvement, release cuts, or regeneration cuts, unless you hire someone to do all the work, you'll use a chainsaw to cut firewood from thinning, to trim trees, and to harvest large trees. You can also use it for clearing trails or hunting blind lanes and even for rough building projects.

A chainsaw is the most important timberland management tool. Make sure you select a saw of a size to fit your needs.

Purchase a good-quality chainsaw, learn how it works, and obey chainsaw safety rules. Chainsaws can be extremely dangerous. According to the Consumer Products Safety Commission, each year nearly one hundred thousand injuries occur involving chainsaws. Following a few basic rules can help you avoid being one of those statistics.

CHOOSE THE PROPER SAW

Select the appropriate size for the chores you expect to perform. Chainsaws are often classified in three general sizes, although they perform most of the same functions.

Light-duty saws, sometimes called "minis," range from 30cc to 40cc and are for use around the home or property on an occasional basis. They're perfect for trimming around tree stands, cutting firewood, and removing storm-damaged trees and brush.

Heavy-duty saws range from 40cc to 60cc and are for use on a regular or daily basis around the farm or ranch, in the woods, or at a job site. These saws make short work of cutting firewood, clearing trees and logs from trails, and even felling some trees.

Professional saws begin at 60cc and go up to 120cc. You'll need a professional saw for felling and trimming trees.

All chainsaws have a suggested bar length for best performance, and saws usually come with an appropriate bar. As a general rule you should select a bar length that is two inches greater than the diameter of the wood you will usually be cutting. This will allow you to make a complete cut through the wood with one pass.

As a general rule, the bar length should be two inches greater than the diameter of the wood you will generally be cutting.

Today's chainsaws come with numerous features—some for ease of use and some for safety.

Purge pump. Provides easier starts.

Decompression valve. Reduces the amount of pressure needed to pull the starter rope.

Bar lubricating system. These are either automatic (the same amount of oil is delivered to the bar all the time), or automatic/adjustable (you can adjust the amount of oil delivered to the bar). Some saws also have a no-oil-at-idle feature, which saves oil and is better for the environment.

Tensioning system. In order to work properly, the chain must be tensioned properly. On older saws, this required loosening the bar's holding nuts and then using a screwdriver to loosen or tighten the tension. On many newer saws the tension screw is on the side where it is easier to reach. Stihl has a feature called the Quick Chain Adjuster that requires no tools to adjust—you simply turn a dial.

Safety Features

- *Chain guard* catches the chain if it comes off while the saw is on.
- *Safety interlock* on some saws uses double triggers to prevent accidental operation of the throttle.
- *Hand guards* keep you from touching the chain if your hand slips off the handle.
- *Front guard* protects your hand if the chain comes off.
- *Chain stop* stops the chain if it comes off.
- *Safety tip* on the end of the bar prevents insertion of the bar end into a cut to help prevent kickback (available mostly on saws marketed to nonprofessionals). Chainsaw kickback can occur when the moving chain contacts a solid object or is pinched near the nose of the bar. The force of the moving chain creates a rotational pull on the saw the direction opposite of that of the chain movement. This flings the bar up and back—in many instances directly back at the operator.
- *Chain brakes* are also used to help prevent kickback. They're available in two forms—manual and inertia. Manual brakes are located in the front guard and are operated by hand. Inertia brakes operate automatically from the force of the saw being thrown back, but they can also be applied manually. Saws equipped with inertia brakes are safer than those with manual brakes. Stihl offers one model with a three-way chain brake. It can be manually activated in the front guard, with the interlock on the handle, or by inertia.
- *Low-kickback chain* can reduce kickback intensity as much as seventy-five percent. Most nonprofessional saws today come with a low-kickback chain.

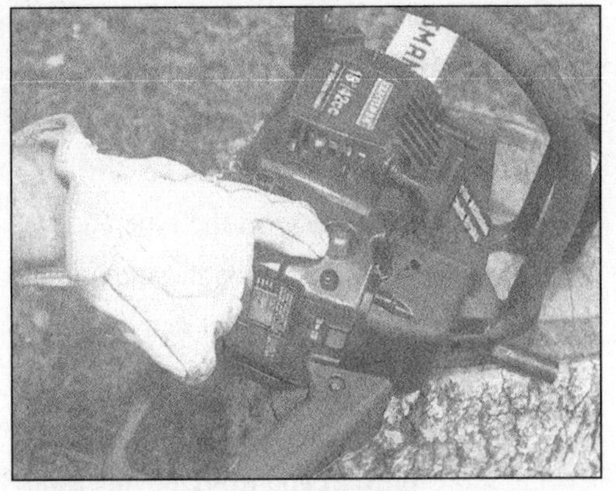

Today's saws have numerous features for ease of use and safety. Fuel bulb primers, decompression valves on larger saws, easy chain adjusting, and dual chain brakes are some of the more important features.

Follow Basic Safety Rules

Wear proper apparel. Wear eye and ear protection, a helmet, work gloves, and protective footwear. Protective chaps or pants can provide even more protection. Avoid loose clothing.

Inspect the saw before use. Ensure that the chain brake is clean, that the brake band isn't worn, and that both inertia and manual activation of the chain brake are in working condition. Sharpen and properly tension the chain.

Start the saw safely. It's safest to start a chainsaw on the ground. Be sure nothing is obstructing the guide bar and chain. Place your foot in the rear handle to make sure the saw sits securely on the ground.

Carefully plan your cutting job. Avoid hazards such as dead limbs, electrical lines, roads, and bystanders (work at a safe distance from other people, but never work alone). Evaluate wind direction and the lean of a tree to be felled.

Protect yourself against kickback. Never modify or remove the chain brake. It is designed to reduce the effect of kickback and prevent possible injury.

Manually activate the chain brake when moving from cut to cut.

Maintain a stable, safe stance while cutting.

Don't saw above chest height.

Don't reach far out to cut.

Timberland chainsaw work involves several chores including tree felling.

Basic Cuts

Understanding the basic chainsaw cuts can make a difference in how easily and safely you can complete a cutting chore. The basic cuts are felling, limbing, and bucking.

Felling. Felling a tree can be difficult or easy, depending on the type and size of the tree, its location, and the intended direction of fall.

- *Plan.* Before you start cutting, plan the fall of the tree.
- *Look for dead branches that may fall off the tree* as the tree itself falls.
- *Examine the area around the tree* and remove obstacles that may prevent you from moving away from the tree quickly if you need to. Your escape route should be to the side, not directly behind the tree.
- *Examine wind conditions and the natural lean of the tree* and, if possible, fell accordingly.

- *Use a wooden or plastic wedge* if necessary to help start the tree in the right direction.
- *Use the notch method for trees with a diameter smaller than the length of the chainsaw bar.* Cut and remove a notch in the direction of the fall and one-third the diameter of the trunk. Make a back cut (a cut on the opposite side of the tree) two inches higher than the notch. It's extremely important not to cut all the way through to the notch cut. Leave a two-inch hinge to give you a way to influence the direction the tree falls. If the tree doesn't begin to lean after you make the back cut and there's room behind the saw blade, use a soft plastic or wooden wedge in the back cut to hold the cut open, and then complete cutting to the hinge and either continue cutting to fell the tree or drive in more wedges to force the tree to fall. As soon as the tree begins to lean, remove the saw and move away fast. If necessary, leave the chainsaw behind.

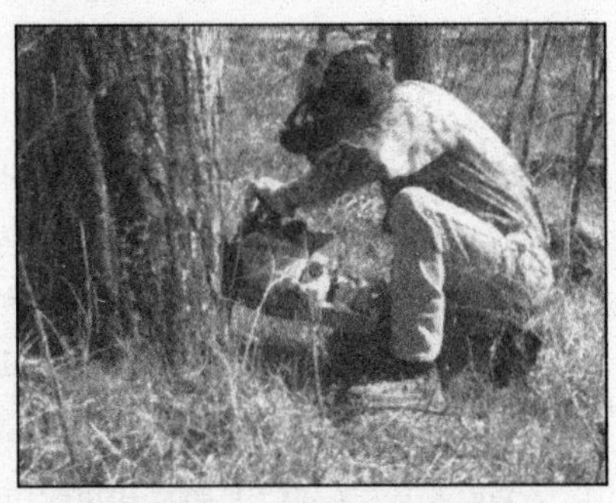

The most common felling method has a notch to direct the fall of the tree.

- *For larger diameter trees, two other methods are illustrated: the apple core and draw across.*

Limbing. Limbing a felled tree can be dangerous because of tension on the limbs caused by the weight of the tree. In some cases it's easiest to limb and buck at the same time.

- *Begin at the outer ends of the upper branches if possible and work back toward the trunk.*

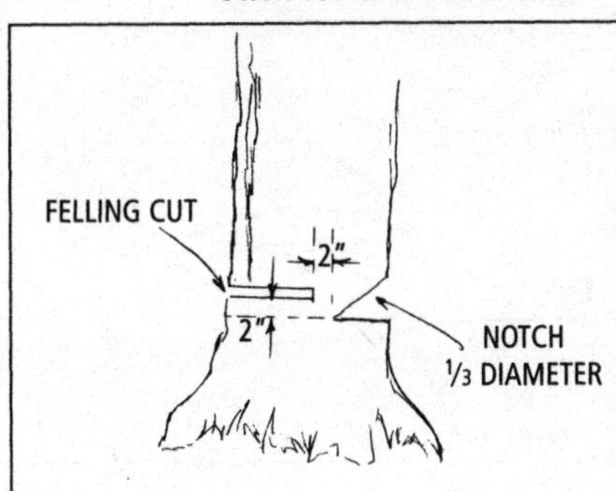

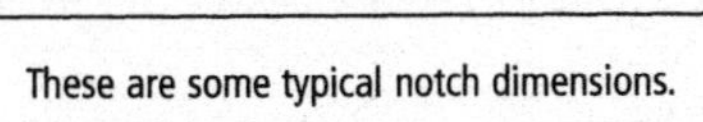
These are some typical notch dimensions.

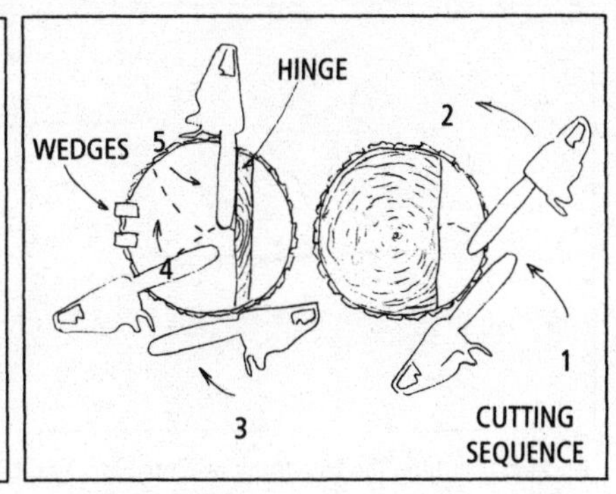

Larger trees can be cut with alternate methods.

- *If the limbs are too high off the ground, cut them off first* and then buck them.
- *If the chain becomes bound, shut off the engine* and use levers to free the saw.
- *Be extremely careful when cutting off limbs the tree is resting on,* because the tree could roll over on you when pressure is released.

The next chore is limbing. This can be dangerous due to the tension on the limbs. Cutting off supporting limbs can cause the tree to roll over on you.

- *Make cuts to release tension from the bottom side of the limb.*

Bucking. Cutting tree limbs into firewood lengths is called bucking.

- *Make cuts from either the top or bottom of the limb,* whichever has the least tension.
- *If you're bucking a large limb or tree trunk,* position small logs beneath it as you cut to keep the log off the ground.
- *Use a cant hook* to turn logs while you're bucking.

OTHER HAND TOOLS

Cruising stick

Twenty-five-foot tape measure

Spray paint for marking trees or branches to be cut. Any bright-colored

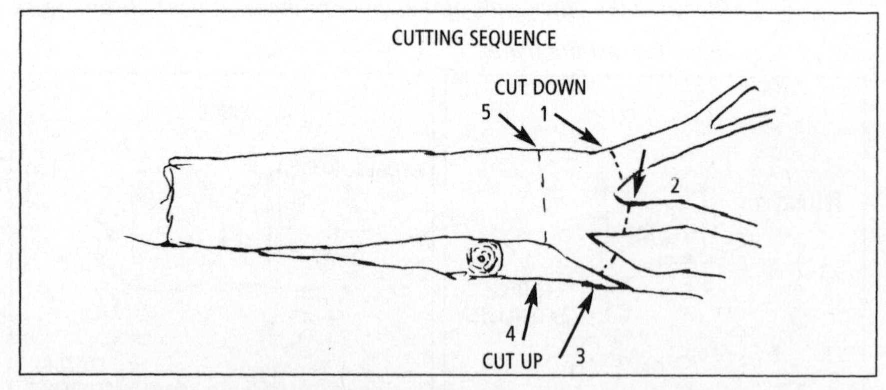

Bucking is cutting the tree trunk or limbs into lengths needed. Use the steps shown to prevent getting the saw chain in a bind caused by the tension on the log.

aerosol spray paint will work. You can also buy marking guns that accept special marking paints from forestry suppliers.

Hand pruner

Pole pruner. Buy one with a combined snip and saw head. Powered pole pruners will make your job easier.

Cant hook for rolling logs

Pry bar for shifting ends of logs

ATV Tools

Although many hunters use ATVs primarily for getting to hunting spots and toting game back out, one of their most useful purposes is helping to create good hunting spots. Carry the following tools on your ATV when you're maintaining your woodlands:

ATVs have become increasingly popular as woodland management tools. With a sprayer and selective herbicide, ATVs can be used to control excess brush.

Winch

Pair of logging tongs

Heavy-duty log chain

With special equipment, ATVs can be used for:

Tank spraying. A tank sprayer behind or on an ATV is the best method of applying herbicide to control low-grade brush and hardwood saplings in pine forests—especially important in the South. The same tank filled with water can be used to control prescribed burns in woodlands.

A sprayer tank filled with water is invaluable for fighting fires.
COURTESY ATV COUNTRY

Holding tools. Tool holders are available to carry shovels, rakes, axes, chainsaws, and other fire-control tools.

Transporting you to your site.

Chainsawing and timber felling. Carry hard cargo boxes on ATVs to hold chainsaws, gas, oil, helmets, and other felling and logging tools. If the felled timber is to be used as firewood, which is often the case in thinning operations, use the ATV to tow a log splitter to the felled trees and an ATV with a trailer to haul out the finished product.

A number of tool holder accessories are also available for toting chainsaws, shovels, rakes, and other fire-fighting and management tools. COURTESY CDS

Mowing. Keeping clearings open is easy with an ATV and a pull-behind rotary mower like those from Swisher and Weekend Warrior.

ATVs or utility vehicles can also be used to skid logs out of the woods.

Skidding logs. Skidding felled logs out of the timber to a loading yard can be done with ATVs using specialized skidding equipment. In fact, an ATV is the tool of choice in areas where low-impact skidding is desired.

- *Skidding cone.* This device is suitable for logs up to twenty inches in diameter. It helps you to avoid obstacles in the log's path, making it easier to skid the log, protecting younger trees, and preventing soil damage. A skidding cone fitted with a Kevlar cable, corner block, polyester tree sling, and capstan rope winch can be used to pull logs out of places inaccessible to ATVs.
- *Logging arch.* Use it to hoist the front end of the log up off the ground, making it easier to pull it and creating less environmental damage.
- *Forestry trailer.* NovaJack's model for use with all ATV models was developed for private foresters and allows loading and unloading of logs by one person. Its protective grating prevents

logs from slipping towards the ATV while transporting and a spring-loaded pivoting tow bar helps start heavy loads and absorbs shock from the trailer. The spring-loaded pivoting tow bar improves control on rough ground and prevents wear on the hitch. The trailer comes with a 4-by-6½-foot box which when installed on the trailer can be used as a transporter for your ATV when you're on the highway. The trailer has a capacity of 3,500 pounds off-road, 1,500 pounds on-road and will haul logs from four to sixteen feet in length.

- *Hardened steel chains* with studs on every second link. These improve an ATV's pulling and braking power.

Another option to using an ATV is the Jonsered Iron Horse log and utility transporter. A small walk-with crawler, the unit goes anywhere you walk. The unit is a gasoline powered "walk-behind" unit on tracks like a "mini" dozer.

The Jonsered Iron Horse log and utility transporter is also a low-damage, easy-to operate, one-man skidding machine. COURTESY JONSERED

Tractor

If you have a farm tractor, even a small twenty hp or so model, you can use it for many timberland management chores:

Spraying. The same sprayer as the ATV sprayer works on a tractor.

Transporting. A tractor's rear platform lifts can be used to transport firewood and tools.

Mowing. Fitted with a rotary mower, a tractor can keep clearings open.

Lifting and moving. The most valuable accessory for a tractor is a front bucket loader, which can be used to push brush into piles, lift log ends for transporting, even hoist deer for dressing and skinning.

Payeur Inc. markets several small- to mid-sized loaders and trailers. Ten models are available for tractors from nineteen to forty-five hp, from eight feet, six inches to fifteen feet, six inches.

A log-splitter, such as the Swisher model shown, and a Kawasaki Mule make quick and easy work of splitting logs and transporting firewood out of the woodlands.

Splitting and Milling Tools

Hand-held splitting maul and wedges. Can be used to split log sections into firewood-sized pieces.

Powered log splitter. Makes splitting easier and much quicker. Numerous models are available, including pull-behind models that can be towed behind a pickup, ATV, or tractor. Three-point hitch, hydraulically powered, tractor-mounted models are also available; they use the tractor's hydraulic system.

Lumber mills. Traditionally, landowners have sold valuable timber to buyers as saw logs. Frequently, however, you can make more profit by milling your own logs. Not only can you use the milled lumber for projects around your property, such as building hunting lodges and sheds, but you can also often use cull logs that are too big for firewood but too crooked or inferior to sell. Gunstocks, gun grips and furniture can be made from these smaller milled pieces that would ordinarily be discarded. To mill lumber, you may choose a chainsaw mill or a bandsaw mill.

- *Chainsaw mill.* Use it to slab medium-diameter logs up to eighteen inches; it is a slow process.
- *Bandsaw mill.* A bandsaw mill can be towed behind an ATV, truck, or tractor directly to the log. This prevents damage to the forest environment caused by dragging or skidding the log to a mill or loading site. TimberKing offers several bandsaw mill models, from big industrial mills down to the 1220, which is suitable for people with just a few trees to mill. The standard 1220 handles logs up to twelve feet long and twenty-nine inches in diameter, but bolt-on extensions allow you to saw logs of any length. The unit features a fifteen-hp electric-start engine and an integral throttle-clutch. A double-crank system controls the height of the cutting head; the second crank moves the cutting head through the log. The 1/32-inch blade removes and wastes very little of the log and produces a surface smooth enough to use for rough projects without planing. Other manufacturers, such as Jonsered, also produce bandsaw mills.

A bandsaw mill, such as the TimberKing model shown, makes it easy to cut lumber from your woodlands. The mill is extremely easy and efficient to operate.

CHAPTER 15

Assessing the Land for Soft Mast

Mast—acorns, and lots of them. That's the key to a big population of white-tailed deer, turkeys, and other wildlife, right? Well, maybe. Both of these popular game species, as well as many other wildlife species, are opportunistic foragers. They eat a wide variety of foods because they have to, and acorns are available for a very short time each year. What do these animals eat the rest of the year? Deer are browsers. In a natural situation, they take a nip here, a bud there, a flower here, a mushroom there, forbs here, and grasses there. Turkeys aren't much choosier. If it tastes good, they'll go for it. Turkeys, however, aren't pure vegetarians. They like meat, too, and insects are a major part of their diet. This is especially true early in the spring when insect numbers are high and the poults have just hatched, are growing fast, and require a lot of food.

If you provide animals a smorgasbord, including soft mast, forbs, grasses, shrubs, and "brush," you'll attract and feed more of them and find more diverse species visiting your land. In many instances all you have to do is protect and enhance species already growing in the area, but you can also propagate plants that will attract more wildlife.

Soft Mast Trees

Soft mast trees produce fruits or seeds other than acorns, which are considered hard mast. At some times of the year soft mast plants provide all the food deer, turkeys, and other wildlife need. If you want to keep turkeys or deer on your property year-round, which often means they are still there when hunting season starts, providing soft mast is important.

A lot of trees fit into the soft mast category, but some are especially important:

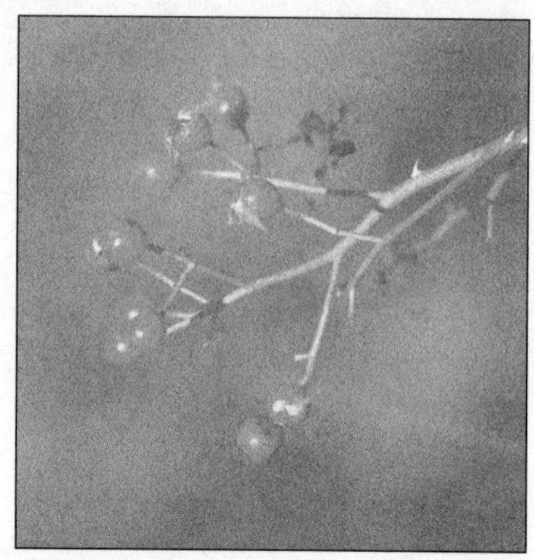

Soft-mast is also an extremely important wildlife food source. Soft mast provides a "smorgasbord" of foods at times of the year when hard mast is not available.

Apple or pear. Hunting white-tailed deer around old orchards is a tradition in many parts of the country and it's still an extremely effective tactic. You can't beat a single old apple or pear tree in the middle of a long-forgotten homestead for attracting wildlife—that is, if the tree still produces fruit. If it doesn't, there are some things you can do to help it produce again (see "Care of Soft Mast" later in this chapter).

Wild crab apple. They often produce more fruit than regular apple trees.

Mulberry. This is one of the most important and easiest-to-grow soft mast trees. In fact, many state wildlife agencies give mulberry trees away or offer them at a very low cost. Producing fruit during late May and into early June, mulberries are a magnet for bushytails. Then the fruit drops to the ground by the bushel and everything that walks, crawls or slithers loves them.

Soft mast includes the many wild fruits, such as mulberries or persimmons.

Persimmon. My great-grandmother thought persimmons and possums belonged together, not only in the woods but on the plate as well. They do, but possums are not the only wild critters that feast off the sweetest and bitterest of wild fruits. Persimmons are greatly loved by many wildlife species, as anyone who walks the woods with an eye to scat can attest. Persimmons grow almost anywhere and reproduce easily.

Dogwood. Known for its beauty as an ornamental tree, the dogwood is also an extremely important soft mast tree. Its fruit are relished by all types of wildlife. During times when quantities of hard mast are low, the soft mast of dogwood becomes even more important.

Hawthorn. A member of the rose family, hawthorns produce fruits (rose hips) that last into the winter and provide food when it's needed most.

Plum. Plum thickets provide not only soft mast but also cover for turkeys, quail, pheasants, and small game. In many parts of the West and Midwest, when you find a plum thicket you can almost count on finding quail or pheasants.

Hackberry. The fruits of the hackberry stay on most of the winter and are relished by turkeys and songbirds.

Osage orange. Squirrels are about the only wildlife that eats the seeds from the fruits of the Osage orange or "hedge" tree. These trees spread rapidly, but offer lots of cover as well.

Pawpaw. This unusual member of a tropical plant family grows in fertile, moist soils. The fruit is shaped like a stubby banana and starts out green but turns purple when ripe. It tastes somewhat like a banana as well. Turkeys, deer, and many other types of wildlife and birds relish the fruits.

Black cherry. The fruits are small and black and grow on a pendulous stem. They're bitter but make great wine and are relished by many types of wildlife when they ripen in late summer.

Black gum. Bottomland trees that produce blue, plum-like fruits with a flattened shape. They're eaten by deer, turkeys, waterfowl, ruffed grouse, and other wild game animals.

Aspen. One of the most important trees in northern forests. All types of wildlife eat its twigs, buds, and even its bark.

SHRUBS

A number of shrubs, from short to tall, also produce soft mast relished by wildlife.

Viburnum. These include the black haw with its small black berries.

Native honeysuckle. These grow as both small shrubs and vines.

Coralberry or "buckbrush." The berries of this small fruiting shrub are readily eaten by deer and turkeys.

Inkberry

Buckthorn

Deciduous holly (also called possum haw)

Juneberry (also called sarvis or serviceberry)

A number of shrubs and small trees also provide soft mast fruits or seeds.

Choke cherry
Witch-hazel
Moonseed
Hazelnut
Alder
Bayberry
Greenbrier
Juniper

BERRIES

Wild raspberries grow rampant on our place. I've often picked berries for a few moments and then surprised a wild turkey foraging as well. And I can also almost count on a fawn bursting from a patch—it's happened many years in a row.

Blackberries, blueberries, mountain cranberries, huckleberries, gooseberries, elderberries, currants, wild strawberries, and bearberries (small, roundish, mealy-tasting fruits eaten by grouse and bears) provide a feast not only for wildlife but for the astute human forager as well. More than one turkey has had its crop stuffed with wild currants when I dressed it!

Wild berries, including blackberries, blueberries, currants, gooseberries, and bearberries are important soft mast foods.

OTHER SOFT MAST

Wild grapes. A wide variety of wild grapes grows in much of the country.

Staghorn sumac. With its fuzzy fall berries, it's another very important source of soft mast.

Wild rose hips. They're very important because they stay on the plant all winter.

One method of increasing soft mast is fertilizing the plants. Scotts Native Plant Fertilizer can be used, as can locally purchased granular fertilizer.

Prickly pear. Although considered primarily a plant of the Southwest, prickly pear is the only widespread eastern cactus. Most cactus are found in the west. Prickly pear is found in the eastern half of the United States. It grows in much of the Midwest and South and is favored by deer, providing food in areas where other mast often does not exist.

CHAPTER 16

SOFT MAST MANAGEMENT

Soft mast can be maintained and planted, but first you must determine what soft mast species already exist on your property and determine sites that are suitable for various species. Some will grow on droughty soils, some on moist soils.

ASSESSING YOUR PROPERTY

Walk your property with field identification handbooks for trees, shrubs, and vines. Mark locations and quantities of the important wildlife food species on a property map, and then decide where you wish to add more plants.

MAINTAINING EXISTING SOFT MAST PLANTS

Many soft mast plants are extremely hardy and need little care; in fact, some may spread to the point of being a nuisance. Some will do better with some yearly maintenance. A good book on garden shrubs, berries, and fruit trees will provide information on how best to propagate these plants or increase their productivity. Here's some general information:

Fertilizing. Fertilizing soft mast plants can often greatly increase production. The standard method is to apply 13-13-13 agricultural fertilizer in the spring followed by an application of 34-0-0 forty-five to sixty days later. Spread about a cup of fertilizer around each plant, making sure it doesn't touch the plant or plant bark.

A simpler method is to use Scotts Native Plant Fertilizer for native browse such as shrubs and vines. With a formula of 36-3-7, each thirty-two-pound bag treats eight thousand square feet, and since it's a timed-release product, you have to apply it only once.

Most wildlife experts agree that white-tailed deer need food sources with crude protein levels of sixteen percent or higher and that native plants like honeysuckle found in the wild normally have protein levels of about eleven percent. This level can be increased to above sixteen percent with a good fertilizer program.

Pruning. The judicious pruning of dead canes and vines can improve the productivity of many plants. Old orchard trees on overgrown homesteads often become unproductive because of neglect. Many of them can be made productive again with a little maintenance. In fact, you'll be amazed at how much some of these old-timers can produce with just a little TLC. Prune the tree heavily with hand pruners, cutting off suckers and topping out the upper limbs with a pole pruner. Cut off dead or diseased branches. Then add a handful of Scotts Tree Tablets around the circumference of the drip line. You might end up with a really productive spot—for both whitetails and yourself.

Adding Soft Mast Plants to Your Property

Most of the plants mentioned in this chapter occur naturally, but you can also plant or propagate these species to increase the variety of wildlife foods on your property.

Buying new plants. Many states offer wildlife bundles that include some of these species at a fairly low price. The National Wild Turkey Foundation also sells some of these species, and private nurseries offer many of the plants as "ornamentals." All of them make great plants for buffering woodland edges, and they can even attract wildlife into your backyard.

Propagating new plants from old. You can propagate many of these trees, shrubs, vines, and brambles.

- *Layering.* Brambles and many berry plants can be propagated fairly easily by layering. Pick a one-year-old stem that can be bent down to the ground without breaking and pin it to the ground with a U-shaped wire. Making an inch-long slit in the underside of the stem where it touches the ground and propping the slit open with a small sliver of wood can enhance rooting, or you can cover the area with soil. To propagate plants to move to another area, simply

Growing lush, thick clover like this requires plenty of lime and fertilizer. You will also have to be lucky to have an adequate amount of rain during the growing period, too. Clover will attract all types of wildlife including deer and turkey.

To clear large areas of brush or forest, it is wise to hire a dozer and operator. Not only will they get the job done more quickly and efficiently, but they will also be able to grade the land so water doesn't accumulate on the plot and kill the plantings.

Smaller-scale food plot planting can be done from the back of an ATV or lawn mower. The Plotmaster has the unique ability to disc, plow, cultivate, plant, and cover in a single pass.

By fertilizing food plots and indigenous plants regularly, you'll attract more game to your hunting area. A spreader attached to an ATV makes the job quick and easy.

Some seeds must be compacted after being planted. A heavy roller does the job.

As crops begin to grow, it is a wise idea to check their progress. By catching problems early (such as bug infestation), they can be corrected and the crop saved.

Planting food plots in clearings that were formerly woodlots can be excellent deer hunting areas. (Note the treestand in the background.)

Food plots can be extremely important wildlife habitat for any number of species.

In states where it is legal to use feeders, gravity units are easy to put up and require less maintenance.

Many seed manufacturers make blends that do not require tilling and can be planted by hand. This planter is over-seeding a plot with a legume in late winter.

Many hunters do not fertilize and prune natural vegetation. This is a mistake. By regularly doing so, you can provide more abundant and sweeter soft mast like this patch of berries. The berries will also last longer throughout the season.

A few trees left standing in a food plot will make deer (especially bucks) feel more secure when leaving the woods to feed in the clearings.

To improve your woodlot, selectively harvest the trees. Prices vary with size and species as well as the quality of the tree. Talk with a forester for advice before cutting any trees on your land.

One way to determine the type of bucks living and feeding in your food plots and woodlands is to use stealth cameras. There are a wide variety of cameras available through sportsman catalogs like Cabela's, Bass Pro Shops, and Gander Mountain.

Planted clearings provide a wide variety of food for wildlife including deer, bears, turkeys, pheasants, squirrels and rabbits. These plots remain wildlife magnets for many seasons when planted correctly and regularly limed and fertilized.

bend the stem down to a pot of dirt. After the new plant is rooted, clip the old stem and move the new plant to the new location.

- *Cuttings.* Making cuttings is the technique used to propagate many trees and shrubs. This method requires more time and effort than layering, but it allows you to propagate numerous plants from a single parent plant. Detailed information is available in the plant propagation sections of many gardening books.

Planting. Whether you buy new plants or propagate them from old, plant them with the same care and attention you would use for those in your yard and garden. Treat each species as you would for garden planting:

- Plant trees and shrubs in holes large enough to contain the roots and allow them to spread.
- Place a bit of compost in the hole before planting.
- Water the hole well before you put the plant in place.
- Position the plant and compact the soil firmly around the roots
- Water the plant well.
- Fertilize. A bit of tree fertilizer such as Scott's Tree Tablets, Nutri-pak, or even a garden fertilizer, can help the new plant get off to a good start.
- Water the new plants until they're off to a good start.

If you're planting fruit trees for soft mast, follow the same procedures as for planting them in a backyard orchard, including the initial pruning at planting time. Although most soft mast is best planted in the springtime, you may prefer fall planting, as the plants don't have to suffer heat and possible summer drought.

My good friend the late Ben Lee not only was one of the country's best woodsmen and deer and turkey hunters, but he could also have been nicknamed "Johnny Honeysuckle." Ben collected honeysuckle seeds in paper bags and regularly scattered them over clear-cuts or any other place he could find where they might grow. He showed me several of his honeysuckle thickets, and we took some good bucks from them. Don't forget soft mast when managing your property to attract more wildlife.

CHAPTER 17

ESSENTIAL VITAMINS AND MINERALS FOR WILDLIFE

I wouldn't think of raising beef cattle without providing free-choice minerals in addition to the forage provided in the grasslands they graze, and I've discovered that many species of wildlife frequent the mineral feeders I put out for the cattle. The most common visitors are deer. When we first established a mineral program with our deer herd, we simply used No. 1 salt-mix, the same material we purchased at the local feed store for our cattle. Several of these initial mineral licks are almost twenty years old, and we keep them active with new minerals each year. As more companies began to offer quality mineral mixes designed specifically for deer, we began to use them in our mineral

Mineral supplements, where legal, can be an important factor in deer herd management. Minerals can supplement natural foods and foods from food plots for larger-racked bucks as well as healthier does and fawns.

program as well, and we've seen great improvement in our deer herd, especially in antler mass.

Many people today consider providing minerals (where it's legal) a major factor in deer herd management. I'm not referring to the flavored and aromatic attractants people put out in the fall in the hopes of luring a buck in close, but mineral supplements that are provided all year. I've experimented with almost every supplement on the market. Some are little more than salt with added attractants. Read the label to make sure you're getting not only calcium and phosphorus but vitamins as well.

Provide Minerals According to Animals' Needs

"Deer start growing their antlers, especially the farther north you go, at the worst time of the year," said Steve Scott of Whitetail Institute. "Snow and ice are still on the ground; things haven't greened up. Research has shown that if there is stunted antler growth earlier, there is no compensatory gain down the road. The pregnant does also need the minerals. Most mineral products are hit hard in the spring and summer and the deer back off of them in the fall and winter. Deer nutritional needs change and Mother Nature provides differently at that time of the year."

Grant Woods, wildlife biologist and consultant, thinks the biggest shortfall of mineral programs is putting the same minerals out all year. "They are not following the deer's physiological cycle," said Woods of landowners who do this. "The physiological cycle is super easy to understand. In the spring when we have green-up, the plants are at the highest in moisture content, and at that time deer crave and need a very high amount of salt. As the summer progresses, depending on rainfall, the moisture content and vegetation decrease significantly. Deer need less salt and a more concentrated source of minerals. Calcium and phosphorus are the two keys, but a lot of other stuff is also involved. Calcium and phosphorus, however, are universal; other minerals are more site-specific. If you take care of the calcium and phosphorus, you're taking care of eighty percent of the need. I start off with a mixture of eighty percent salt and twenty percent 'good stuff' in the spring. This serves two purposes. Deer have to have salt. As the season progresses, say in early June, I change to about 50/50 salt-mineral mixture. Then, in late summer, during what I call the 'leather' stage because the vegetation feels like leather with virtually almost no moisture, I change to eighty percent good stuff and twenty percent salt.

"It's important to keep the mineral flow going into the skeletal system. Unlike humans, deer store calcium and phosphorus in their skeletal system and mobilize it either for antler development or lactation. Think about casting off something the size of our arm or leg and rebuilding that annually."

To address this concept, both Whitetail Institute Cutting Edge supplements and BioLogic Full Potential supplements feature three products in three separate bags to be placed at three times of the year.

Provide Minerals Year-Round

Three factors are involved in producing quality bucks with good racks: age, genetics, and nutrition. Although in many cases you can't easily control the first two, you can quite easily add nutrition, especially minerals. In fact, in some areas, with the proper use of minerals you can go from seeing forkhorns and small basket racks to seeing nicely formed six- and eight-pointers in the first season of application. Minerals can also increase the general health and body size of your deer herd if used properly in a good wildlife management program. Minerals are not a panacea—they must be used in combination with a good management program that includes providing plenty of year-round protein in food plots, managing browse, and managing mast.

Loose minerals and blocks are available in two forms: those specifically designed as mineral and vitamin supplements and those with added "attractants." Some of the latter also contain protein foods as well. Some minerals are also designed to be used in conjunction with attractant protein supplements. A combined program using minerals fed free-choice year-round and mineral/protein blocks available from midwinter through late summer is extremely effective.

A wide variety of minerals, "designed" for deer management, is available these days.

For instance, Pennington Seed suggests using its RackMaster Deer Mineral Salt Lick and RackMaster Deer Mineral/Nutrition Block together. Follow these directions:

First, dig a hole about thirty-six inches square and eighteen inches deep. Mix about a third of the soil thoroughly with a twenty-five-pound bag of RackMaster Mineral/Salt Lick and place this mixture back in the hole. This should fill the hole up to about six inches from the former ground level. Dispose of the remaining dirt by removing it from the area.

Next, remove the cardboard from the RackMaster Deer Mineral Nutrition Block and place it on top of the mixture of dirt and Deer Mineral Salt Lick in the hole, sinking it into the mixture so that about two-thirds of the block is exposed.

Minerals don't do any good—either in attracting deer or providing supplements—if they're applied only a few days before the hunting season. Minerals and vitamins should be fed free-choice year-round. When you're first starting or when you're replenishing mineral licks and blocks, begin in mid-winter to feed the does in your herd. The most important time for bucks is from antler drop to antler hardening time. If you can't feed all year, at least keep licks maintained and fresh from the beginning of February through the first of October (if allowable in your state).

What Minerals to Provide

Phosphorus and calcium make up most of a buck's antler-growth mineral requirements, with small amounts of other minerals and some vitamins also needed. Does also need phosphorus and calcium during pregnancy and for milk production after the fawns are born. In some areas all or some of these minerals may be available naturally, but in most parts of the country they need supplementing.

"A live, growing antler is about forty percent mineral, probably twenty-five percent water, blood, etc., and seventeen to eighteen percent amino acid or protein," said Dr. Kenneth Cromwell, developer of Antler 'n Acorns Magnum Rack Deer Mineral. Cromwell has a doctorate in animal nutrition and also makes traditional bows and is a devout deer hunter. "During the antler-forming stages you have to have a very high quality protein in addition to the minerals," he explains.

Cromwell developed his product after a failed experiment in trying to produce a condensed, compact, strong leg bone for thoroughbred race horses. "Instead of getting a compact bone, we got a lot of real coarse bone growth. I chalked it up as a total failure. A couple of years later I was sitting on a deer stand and I was seeing only small-racked bucks. I got to thinking there had to be some way to enhance antler development—and then I began thinking about that failed project.

"It's not just the calcium and phosphorus in the minerals, it's the amount and the balance or ratio of the calcium to the phosphorus [that's important]. Then there are several trace minerals that also affect whether you are going to get good antler growth. In fact, you might say that the three most important words in nutrition are balance, balance, balance. We looked at about sixty-two nutrients when we formulated a ratio in developing our product."

Another mineral supplement I've tested extensively is Vita-Rack from Hunter's Specialties. Vita-Rack 26 contains a blend of fourteen trace minerals and twelve vitamins.

David Fuhr, Senior Research Scientist at Fuhr Research Labs, who conducted the testing for Vita-Rack 26, said, "Our studies have shown that deer are often lacking in the essential vitamins and minerals they need to maintain good health. Many factors, including weather, geographical location, and current farming practices, can limit nutrients found in the wild. Our goal was to develop a year-round nutritional program that would address those deficiencies in a deer's diet. We have seen incredible improvements in the body size and health of the deer on our test farms. Bucks are growing larger, with heavier racks, and multiple births are common among does in the herd. One unexpected benefit of feeding the supplement has been the huge reduction of ticks and other pests bothering the herd. Our main focus, however, was to produce a product that was driven by performance, not cost."

Some of our mineral licks have been established for a decade or more and still have continuous use.

Landowners frequently use ordinary livestock salt mixes containing trace minerals because animals are attracted to salt and the salt also limits the amount of minerals that can be consumed at one time. These mixes are useful for cattle producers and can be used for deer herds. In most instances, however, although the mixes are economical, they don't provide deer with enough calcium and phosphorus.

It's important to check the "guaranteed analysis" on the mineral mix bag to determine what you're getting. Most mixes are about half salt, as salt is used to attract deer to the minerals. Other ingredients should include, of course, calcium and phosphorus, as well as magnesium, potassium, sulfur, manganese, zinc, iron, copper, iodine, and cobalt. Some mixes also include trace amounts of vitamins such as A and D.

Each manufacturer suggests using its products in a different way.

Imperial Whitetail 30-06 Mineral/Vitamin Supplement. Whitetail Institute of North America suggests putting five to twenty pounds of this mix in the lick when you're first establishing it, but once the lick is established, you should add twenty to one hundred pounds when you replenish it. This helps cut down on human scent by reducing the number of times you have to revisit and replenish the site. They suggest a lick for every ten to forty acres of land, depending on the size of your deer herd and the habitat. Once the licks are in use you can determine whether you need to cut back or add licks.

BioLogic Full Potential Mineral. Mossy Oak's Stan Elliot explains that their mix contains three bags of minerals that are set out in sequence over a three-month period. The first bag is mostly attractant and has the highest percentage of salt. When you get to the third bag, there's less salt and more minerals and vitamins.

Megabucks AGF. Clear the forest litter from a flat spot and spread three to five pounds in the cleared spot. As the minerals are consumed, deer will dig a hole to get at them.

Antler King Trophy Deer Mineral. Developed by animal nutritionist Todd Stittleburg, who suggests dumping the entire contents of one bag near deer trails or other areas frequented by deer. Dump it all in one pile, don't sprinkle it over a large area, and don't offer additional salt or salt-containing products in the area.

It's important to pick a good spot for the feeding station and use the same station year after year. Check the lick once a month to see how much is being

To be effective, mineral licks should be properly located away from roads and property boundaries, but near cover. Follow manufacturer's directions as to application, as they differ.

used and replace with minerals as needed. Antler King's studies have shown that consumption will vary from one to five ounces per head per day. The greatest consumption will be from March through August, the period when most antler growth occurs and fawns are feeding on the does.

Evolved Habitat Block Topper. This attractant can be added to a mineral lick to increase usage. Set out the mineral mix and then pour the powdered portion of Block Topper on top and around it. Next, slowly pour the liquid portion of Block Topper on top and around the minerals and it will candy-coat the entire site.

In addition to mineral supplements, protein supplements are also extremely important. Several companies produce protein blocks or mixes: Hi Protein Big Buck Block by Antler King, Rut Nuts by Megabucks, Antler 'n Acorns, BuckMaster 400, and Imperial 30-06 "Plus Protein" by Whitetail Institute.

"All you can do with nutrition is allow the deer to show their genetic potential," sums up Cromwell. "The greatest products in the world will never build antlers larger than those determined by an animal's genes. But poor nutrition can keep you from seeing the quality genes you may have in a deer you are hunting."

Note: Make sure you understand your state's ruling on the uses of minerals, salt, and protein supplements for deer herds. Some of these may be illegal if used in conjunction with hunting. Some deer blocks containing proteins such as corn chips may be considered baiting for turkeys and may be illegal even if deer season is over.

VITA-RACK 26

THE FUNCTION OF MINERALS IN WILDLIFE AND EFFECTS OF DEFICIENCIES

Major Mineral	Function in Wildlife	Effects of Deficiency
Calcium	Growth of bones and teeth and antlers, function of muscles and nerves	Abnormal bone and antler development, loss of muscle tone
Phosphorus	Growth of bones, teeth, and antlers, energy metabolism and enzyme systems, proper protein utilization	Poor feed efficiency and weight gain, abnormal appetite, lower reproductive performance, abnormal bone and antler development
Potassium	Function of nerves, part of enzyme systems, maintains water and minerals balance	Poor growth and feed efficiency

Major Mineral	Function in Wildlife	Effects of Deficiency
Sulfur	Essential part of some proteins	Poor growth
Sodium	Function of muscles and nerves, maintains water balance	Poor feed efficiency, weight gain, abnormal appetite
Chloride	Forms hydrochloric acid in abomasums, aids in breakdown of protein	Poor feed efficiency
Magnesium	Involved in almost all body processes	Poor feed efficiency and weight gain
Trace Mineral	**Function in Wildlife**	**Effects of Deficiency**
Manganese	Essential for good bone development and feed utilization	Anemia, poor growth and feed efficiency
Copper	Needed for blood and for proper feed utilization	Anemia, poor growth and feed efficiency
Zinc	Influences rate of nutrient absorption	Poor growth
Iodine	Component of thyroid hormone which controls body temperature and rate of metabolism	Poor feed efficiency
Selenium	Needed for growth and reproduction, involved in enzyme system	Poor growth

Trace Mineral	Function in Wildlife	Effects of Deficiency
Cobalt	Necessary component of Vitamin B-12 and enzymes that digest food	Poor growth
Iron	Component of red blood cells	Anemia, poor growth

The Function of Vitamins in Wildlife

Vitamin	Function in Wildlife
Vitamin A	Necessary to support growth, vision, reproduction and involved in bone development and antler growth. Aids in controlling infections.
Vitamin D	Necessary for the mineralization of bone development and antler growth. Also helps maintain proper functioning of muscles, nerves, blood clotting and cell growth.
Vitamin E	Involved in the enzyme system, acting as antioxidants at the cellular level. It also has an important role in selenium metabolism.
Vitamin B-1	A catalyst in carbohydrate metabolism, enabling carbohydrates to release energy. Also acts as a natural insect repellent.
Vitamin B-2	Acts as a critical cofactor or coenzyme in the metabolism of fats, carbohydrates, and amino acids
Vitamin B-3	Is critical to cellular respiration and essential for the metabolism of carbohydrates and fats
Vitamin B-5	Is required in the metabolism of fat, protein and carbohydrates. Plays a major role in many cellular enzymatic processes.

Vitamin	Function in Wildlife
Vitamin B-6	Is required in the synthesis and metabolism of protein and amino acids. It also plays a role in the formation of red blood cells.
Vitamin B-12	Is critical to normal nerve cell activity, DNA replication, and the development of red blood cells
Vitamin C	Functions as a powerful antioxidant, is a key factor in collagen. Helps protect vitamin A, E, and some B vitamins from oxidation.
Biotin	Serves as a critical cofactor in the metabolism of carbohydrates, proteins, and fats. Also aids in antler growth.
Folic Acid	Plays an important role in the synthesis of the genetic material DNA and RNA.

COURTESY HUNTER'S SPECIALTIES VITA-RACK 26.

Another option, according to the folks at Hunter's Specialties, is a mineral/vitamin feeder. Make a sixteen-inch-square box of two-by-six treated lumber placed on edge so the box is six inches deep. Add a treated plywood bottom. Drill a few holes to let rainwater drain through. Fill the box with Vita-Rack 26 and place several boxes along trails.

CHAPTER 18
SUPPLEMENTAL FEEDING

One of the most serious disagreements among those who manage property for wildlife concerns supplemental feeding. You'll find opinions on both sides among biologists, politicians, landowners, and hunters. Some states don't allow supplemental feeding, others have stringent regulations if the food is considered bait, and some allow it unconditionally. The main argument for supplemental feeding is that it provides feed at a time when wildlife need it the most. The main argument against it is that feeding attracts wildlife to specific areas, and a concentration of wildlife can lead to the possibility of disease, predators and, of course, poachers—and could be considered baiting. Even where feeding is allowed, the quality of feed is a concern. A few years back the Arkansas Fish and Game Department issued an alert because "deer corn," a common feed that is often rejected by mills for livestock feed, may have been infected with alpha toxin, a fungus that can grow on corn. Alpha toxin can attack the liver, lungs, kidneys, and other organs of humans as well as domestic and wild animals. "Piling up corn for deer and other wildlife promotes the growth of the fungus involved in alpha toxin," said Dr. Dave Urbston of Arkansas Game and Fish Commission's Wildlife Management Division. "If the corn kernels are scattered, the growth of the fungus is diminished, but there is still danger."

Whether or not to feed also depends on the availability of other foods. If you're in an agricultural community where corn and soybeans are the main crops, putting out corn may not be of much use—at least it wouldn't have been years ago. These days, with clean cropping methods, by midwinter there's often little food left on the ground. From my own experiences over the past thirty years of managing a Missouri farm for wildlife and cattle, I've found supplemental

In some instances, supplemental feeding, where legal, can also be an important part of an overall wildlife management plan. COURTESY MOULTRIE FEEDERS

feeding to be an important part of my management program. After all, I supplementally feed my cattle—why not the wildlife as well? My cow-calf herd is fenced off from the managed wildlife areas such as timber tracts with clearings, and the entire property is interspersed with both legume or clover food plots and annual food plots of soybeans, milo, wheat, and oats. My neighbor runs a dairy operation with large corn fields followed by winter wheat. The deer tend to stay year-round, moving from my property to his and back.

Turkeys are another matter. One year I didn't start the supplemental feeding program as usual just after deer season or when we quit hunting in early winter. I was too busy traveling and writing and didn't get the feeders out. Usually a couple of hen/poult turkey flocks and a gobbler flock have wintered, but it didn't happen that winter. A neighbor down the creek complained at the coffee shop about the tremendous number of turkeys coming to the feed he was spreading for his cattle. "Several hundred," he complained, and they were tearing the covers off his hay bales. That winter was really tough where I live. I also noted fewer quail, even though I had begun an extensive natural prairie-restoration program sponsored by Quail Unlimited that was intended to bring quail back by planting warm-season grasses in some areas.

It's simple: Supplemental winter food should, in some instances, be provided for wildlife.

Types of Feed

Corn. I've traditionally used corn—the same corn I feed my calves for fattening out. Corn is very common, it's the most economical feed, and both deer and wild turkeys like it. I do not, however, use the cheaper "deer" corn.

Morgan Richardson of Sheridan, Arkansas, a wildlife biologist with International Paper Company said, "Our recommendation to our deer clubs is to use only certified field corn. Just because corn is in a camouflage bag doesn't mean it's certified. Read the label."

Soybeans. My good friend Dan Moultrie, however, has convinced me that soybeans are a better bet. He's the president of Moultrie Feeders, an automatic feeder business. "Corn is deer candy," said Dan. "It's extremely palatable, and they'll come to it over almost everything else. Corn has about seven to eight percent digestible protein, and it's high in carbohydrates. Some folks say corn isn't good for deer, but it's actually good energy food; in fact, it's identical to acorns.

"If you want to grow big-racked bucks, however, you need to pump in more protein. Since deer are browsers and don't tend to stay in one spot and eat, you need to concentrate the protein. Soybeans run up to forty-five percent digestible protein. A few bites of soybeans is like giving steroids. Soybeans are all I feed these days, but you do have to train deer to come to them. When you start feeding, use a 50-50 blend of soybeans and corn. The deer will pick up the corn first every time, and then they'll start eating the beans. Then go to seventy-five percent beans and twenty-five percent corn, and then all beans. Once you get them converted, they come just as well as they did with corn."

Commercial feeds. These are available, although I haven't experimented with them. Larry W. Varner, a wildlife consultant with Purina Mills, has done extensive studies on supplemental feeding. In his experiences in south Texas, he's found that supplemental feeding is very important during drought years and can have an important impact on the makeup of a deer herd—and in particular, on trophy bucks. Varner suggests that landowners create food plots before putting out food. "Spring food plots have the greatest effect on antler development and fawn growth. Fall plots impact body weights and winter survival. But if a drought occurs, those foods and the natural forbs and grasses that deer rely on become increasingly scarce."

The biggest problem I've found with commercial pelleted food, other than getting deer trained to eat it, is that the feeders have to be absolutely dry. Once the pellets pick up moisture, they disintegrate. Varner suggests using corn to train the deer to come to the pellets. "Locate feeding areas in areas of high deer activity and minimal human activity," said Varner. "Near water, along creek drainage, or along a game trail are good places to start. It is critical that deer first be attracted to the area where the feeding pen is located. The quickest way to accomplish this is to use a timed feeder that slings out corn twice a day. If that is not possible, then spread corn by hand daily."

Once the deer are coming to the corn, switch to pelleted foods. They provide the proper balance of high-quality protein, vitamins, minerals and other nutrients deer need, and pelleted formulations prevent sorting and nutrient imbalances common with grain or grain mixes and feeding programs designed for other species. These companies provide pelleted foods:

Purina Mills. Because of differing nutrient requirements for the different seasons, Purina offers three products: High Protein Deer Checkers for January to August; Deer Checkers sixteen percent for August to September; and High

Energy Deer Checkers fourteen percent for September to December. These are regionally formulated to help overcome geographical as well as seasonal differences in forage nutrient content.

Megabucks. Rut Nuts pellets from Megabucks are thirty percent protein and are designed to be mixed with corn and to be fed free-choice. By varying the amount of corn and pellets, you can vary the amount of protein. Twelve percent protein is suggested for winter and twenty percent at the time of antler drop to obtain maximum trophy antlers.

Pennington Seed Company. Deer Acorns are specially formulated to simulate real white oak acorns and can be used in feeders or scattered.

Winchester. Deer Corn is cleaned and certified safe from alpha toxin contamination.

Kenco. Quick Draw Feed Supercharger is a feed enhancement formulated of peanut byproducts with a nutty aroma that attracts deer. It treats up to a thousand pounds of corn.

Supplemental feeding will not take the place of a good management program that includes providing or enhancing native food plants as well as food plots. Supplemental feeding can, however, be an important part of a total management program for what I consider the big three—deer, turkeys, and quail.

CHAPTER 19
GAME FEEDERS

The use of game feeders continues to be a controversy and feeding, especially baiting for hunting, is not allowed in some states. Where they're allowed, game feeders can attract deer to your property or hunting lease and improve the overall deer herd health. They do, however, need to be used properly and in conjunction with other deer management methods. Since proper amounts of protein are necessary for good antler growth, the right kinds of feeds at the proper times of the year can be a factor in increasing antler mass in bucks.

TYPES OF FEEDERS

Game feeders are available in three forms: *automatic throw, free-feed* and *gravity feed.*

AUTOMATIC FEEDERS

Automatic feeders use an electric motor with a spinning wheel to dispense feed. Animals, including deer and turkeys, quickly become trained to the sound of the feeder spinning. In fact, Bill Harper, former owner of Lohman Game Call Co., joked several years ago that deer and turkey calls that mimicked the sound of an automatic feeder would work very well for hunting in Texas. Automatic feeders use either a photocell to turn the motor on and off at dawn and dusk or timers that can be set to turn it on and off at specific times.

Photocell feeders. "Photocell feeders are the easiest, best way to get started," said Dan Moultrie of Moultrie Feeders. "They're usually programmed to come on about an hour after daylight and an hour before dark. They automatically set themselves every day. Daylight doesn't stay consistent with time

In most instances, feed is distributed by feeders. Automatic feeders with spinning wheels are extremely effective, but do take some maintenance.

of day, but most hunters hunt by daylight and darkness anyhow, so the photocell feeders follow hunting habits. And deer learn to move during that hour after daylight and hour before dark. They know that if they're not there to pick up the food at that time some other animal will, so it encourages non-nocturnal feeding at the time most hunters are in the woods."

Timer feeders. Timer feeders can be set to dispense feed any time, and if you set them to dispense several times a day, the amount of food available is increased and more animals can visit. Often dominant animals keep others away from a single feeder that goes off only twice a day. In fact, turkeys sometimes run deer away from feeders!

"Timer feeders are for those who have been feeding for some time and are willing to work at it more," said Moultrie. "For instance, one trick in Texas is to set the feeder to go off in thirty-minute sequences to attract deer that might not be in the area when it first goes off, but may move through later. The main drawback is that if you set it to dispense at 5:00 in the afternoon, part of the year that's perfect, but part of the year you're feeding after dark. A timer takes consistent monitoring. On the other hand, instead of waiting on light level triggering, you look at your watch and know that it's 5:00, so the feeder is going to go off."

Both automatic and self feeders (or free-choice feeders) are available in several sizes, ranging from five-gallon-bucket size to units capable of holding up to two thousand pounds of feed. The larger sizes are available with either fifty-five-gallon barrels or preformed metal hoppers. Smaller feeders usually hang, while larger ones sit on free-standing stands. The smaller (thirty-pound capacity) feeders will usually hold enough for about a week of twice-daily feedings.

Although barrel feeders are the most economical, they do have some problems that I've discovered over the years. One is that the feed in the bottom doesn't always shift down and eventually becomes moldy. Internal barrel funnels are available to solve this problem. Another problem is that some smaller bucket-style feeders use a snap-down lid with tabs. The lids are hard to get off and on without taking down the feeder, and they often allow rainwater to get in. The larger barrels use a bolt ring lid that allows water to collect on top of the feeder. Moultrie Rain Caps are held in place with the existing lid ring, but they're shaped to keep water from collecting on top of the fifty-five- and thirty-gallon drums. A special indentation accepts a Moultrie Feeder Solar panel.

EXAMPLES OF AUTOMATIC FEEDERS:

Sweeney. Over thirty years ago John Sweeney perfected the first programmable deer feeder. Many units that were introduced thirty years ago are still in the field in active use. Sweeney's patented Directional Feeders with one hundred- and three hundred-pound capacities are available in hunter green.

Kenco. The Tornado Feeder is a formed poly-barrel that won't rust. It has a windproof, easy-off lid with no bolt ring; quick-mount legs that require no mounting rings or hardware; and a see-through area that allows you to check the feed level. The Kenco Smart Timer Control is self-programming and allows you to drive one, two, or three motors to create a specific throwing pattern.

B.A. Products. Winchester Automatic Feeders use a square galvanized steel container that is weather-tight and has a snap-lock top. Winchester offer digital and quartz timer units.

San Angelo/All-Luminum Products. Dawn and Dusk Feeders use a six-volt battery to broadcast up to forty feet. You provide the barrel.

Automatic feeders use D-cell, six-, or twelve-volt batteries. Gel-cell rechargeable batteries can also be used, and most companies offer rechargeable batteries, trickle chargers, and solar panels that can be attached directly to the feeder. The Chaslyn Company offers a six- or twelve-volt solar charger, the Sun Maxx, that I've tested on feeders. It comes with alligator clips that are simply fastened to the battery connectors of a six-volt feeder motor, and no batteries are needed.

One of the problems with feeders is getting feed into them. Hanging feeders often must be lowered unless you want to stand on tiptoe in the back of your pickup trying fill them. Another problem is that sometimes there just isn't a tree limb where you can hang the feeder, but tripod-style feeders solve this problem:

Game Country. Their Heavy-Duty Tripod System features a winch and pulley that allow you to lower the feeder on galvanized steel aircraft cable. Game Country also offers varmint guards for their feeders that keep raccoons and squirrels from getting to the feed.

Moultrie. This company's feeders range from a portable 6½-gallon hanging model to the Premium Magnum Feeder, which holds 220 pounds of corn and can be set to feed up to four times per day. A window provides easy viewing of the remaining feed, and the quad-leg design is very stable. Varmint guards are available from Moultrie to mount on any drum or bucket.

American Hunter. They offer feeders ranging in size from a one hundred-pound to two thousand-pound capacity. The Easy Loader includes a ladder for ease in filling.

FREE-CHOICE FEEDERS

Free-choice or free-feed feeders are available as simple troughs or gravity drop dispensers. They eliminate the hassle of batteries and offer feed free-choice at all times. "Free-choice feeders have advantages and disadvantages," said Moultrie. "The biggest problem is that free-choice feeders encourage nocturnal feeding habits, but there's food

Gravity drop feeders are extremely simple, require less effort in use, but do not feed as many animals at the same time.

available any time and in any quantity they want. Another problem with free-choice is that rodents and other animals get in it and put their droppings in it and it's really not a healthy situation unless the dispensers are cleaned pretty regularly. And a feedlot situation tends to develop around them because the animals tend to stand in one spot." Examples of free-choice feeders:

Knight and Hale. I've tested this gravity-drop feeder, and it's very simple—just a plastic tube with an opening at the bottom and a lid at the top. You tie it to a tree and fill it with feed.

Winchester. The Protein Feeder Station comes in four hundred-, six hundred- and one thousand-pound capacities with a weather-tight hopper. An optional rain shield system helps protect feed. Multiple protein cube sizes can be dispensed, with 1½ pounds of feed generally available at all times. Wild hogs are unable to damage this feeder, it fits into a regular-size pickup bed, and it's also easy to fill from the tailgate of a pickup. With four legs instead of the usual three, it's stable and wind resistant.

American Hunter. American Hunter sells a two thousand-pound free-choice feeder that's made of twenty-gauge galvanized steel and has a window to the inspect feed level. An optional vibration system produces timed feedings. Their Trough Regulated Feeders use a motor to drop feed into a trough below the feed bin. They're easy to move and fill and can be set to give up to 144 feedings a day.

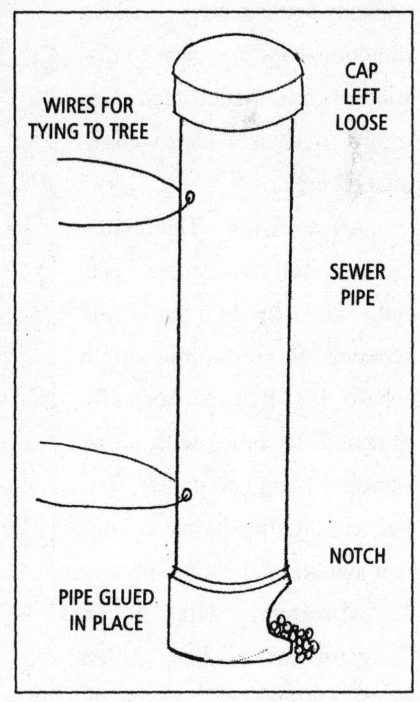

A very economical gravity drop feeder can be made quite easily from sewer pipe and caps.

Kenco. Their thirty-pound-capacity windfeeder is adjustable for any type of pellet or grain. Breeze swings the metal wind vane assembly to trickle out feed, a simple mechanism with no timer. If you hang the windfeeder low, deer will use it as a demand bump feeder as well.

Game Country. Use their Turkey Feeders if you want to attract turkeys as well as deer. They use a gravity-drop design and come with a sixteen-, thirty-five-, or fifty-five-gallon drum. Feed is funneled to the center of the drum, where turkeys have access to it. The feeders are easy to fill, clean, and maintain.

Home-made. You can make very economical gravity-drop feeders from sections of four- or six-inch PVC pipe. Cut the pipe to about three feet in length using an old handsaw. Smooth the cut edges. Glue a cap on the bottom end of the pipe and then cut a small notch in the bottom so the corn can fall out. If the notch is too big, too much corn will fall out at once. Bore a couple of small holes through the pipe at about a foot from the bottom and 2½ feet from the bottom. Thread wire through the holes to fasten the feeder to a tree. Fill the feeder and slip another cap over the top. These feeders are so economical that you can make several.

Tailgate Feeders

Portable tailgate feeders are useful because more feed can be dispensed at one time than with other feeders and it can be scattered so more animals can feed rather than requiring them to group together.

Winchester. These tailgate feeders have eighty- to one hundred-pound capacity and can be used with a directional spinner plate to distribute corn, pellet feeds, or grains over a wide, wedge-shaped area.

Audio Link. The Game Feeders Tailgate Feeder bolts onto either the bumper or the receiver of your trailer hitch where the ball is normally attached. It comes with a one hundred-pound container, funnel, lid, wiring harness, control switch, and hardware.

Moultrie. The Truck Tailgator has a fifteen-gallon drum, mounts into a two-inch box hitch, and conveniently

Tailgate or ATV feeders can be used to scatter grain over wider areas and prevent crowding the wildlife in one spot.

plugs into a cigarette lighter. I've also used Moultrie's ATS ATV feeder on my Kawasaki Prairie 400 for spreading not only feed but also fertilizer and seed for food plots. It works off the ATV's six-volt battery and operated with a hand-held switch.

Choosing a Feeder Location

Regardless of where you place a feeder, deer and other wild game will find it. Picking the proper location, however, is important. Obviously you don't want the feeder to be visible from any road or you'll be inviting road hunters and other unwanted guests. The best locations are on well-drained soil near major game trails and high deer-use areas. Placing feeders near food plots will help provide year-round food and keep deer trained to come to a specific location all year even if you don't keep the feeder in operation year-round.

"You really don't want too many feeders or you scatter the animals out too much," said Ron Ferguson of Winchester, B.A. Products. "For one hundred acres of land, for hunting purposes, I wouldn't put out more than three feeders. You want to keep them ten to fifteen yards from the edge of cover. Otherwise the bigger bucks aren't going to want to go out and feed during the daylight hours. They'll wait until dark. Don't place your stand too close to the feeder, and don't place the feeder right on a game trail."

CHAPTER 20

Monitoring Food Plots, Mineral Licks, and Feeders

Success in food-plot planting is easy to see: it's green and lush or it's not. Determining who visits your food plots, mineral licks, and feeders—and when they visit and how many visit, however, are also important. Monitoring food plots is especially fun if you use monitors that photograph the game.

Game Monitors

Years ago, when I first started working with food plots, I used binoculars to watch fields to determine the primary trails deer used when entering the fields. I then used a garden trowel to rough up the trails and smoothed them down again. Checking the disturbed soil morning and evening gave me an idea of the number of deer using each field. My attempts were crude, and I acquired trail monitoring devices as soon as they were available.

Event Monitors

Trail Timer TT-100. This was the first event monitor available. It uses a digital clock attached to a thread stretched across a trail. When an animal walks into the thread (which is strung over a trail and is usually tied to a small sapling across from the timer), it pulls a switch that stops the clock and records the time and date. The direction of travel is indicated by the direction of the thread break. These units are still available and are inexpensive enough that you can

Trail monitors can be used to monitor food plots, mineral licks, and feeders. Those with cameras allow visual records of the wildlife visiting, as well as the times.

use several to scout edges of a food plot to determine the most used trails as well as the times they're used.

TrailTimer "Plus 5." This was the next step up in technology, and, of course, cost. Using infrared technology to detect the body heat of game, the unit records events on a series of five digital clocks, one at a time, and the information is stored until retrieved. The clocks register time plus AM or PM and the date. They must be checked and reset daily.

TrailTimer "Plus 500" Multiple Event Game Monitor. This device uses infrared technology and records and stores up to five hundred events with date and time. I've tested these units intensively on both food plots and mineral licks. They have a relatively small focus beam—three feet in diameter— which minimizes false signals and a sleep mode that conserves battery life.

TrailMaster Infrared Trail Monitor. I have used these devices for about ten years and have found them to be extremely high-quality units. I've had absolutely no problems during their long usage. The TM550 and TM TrophyTimer are both passive infrared models priced within the budget of most hunter/landowners. They detect both heat and motion in the monitoring area. The sensitivity area forms a sixty-five-foot-deep wedge that spreads out from the monitor for 150 degrees. Each unit can store one thousand events with both time and date. They have "Passcode Protection," a personal code that can be programmed into the monitor to protect the information gathered. The TM550 can be used with all Trailmaster accessories, including a camera, but not with video. Additional controls on the TM550 allow you to set the unit number and control the event delay time. The event delay time allows you to adjust the unit to record events as close together as six seconds or as far apart as two minutes.

TrailMaster also manufacturers the TrailMaster TM1500 and TM1000, which are active infrared units. They are much more costly than the passive models, but they can be programmed to be game-selective. They also offer a computer statistics pack. With this software you can download data and create animal activity charts complete with bar graphs in a variety of colors. These units are a bit more complicated to use and need more theft protection than the passive models.

CAMERAS

TrailTimer "Plus" TT-2000 Camera System. The "Plus 5" or "Plus 500" above can be used with this system not only to record when animals use a certain area, but also to take photos, and that's where the real fun begins. I've been using trail monitors with cameras for a number of years, and getting back a roll of film is always exciting—you never know what you'll see. The TT-2000 Camera System is actually a means of holding a camera (which you provide). The camera must have auto film advance, auto exposure, and auto flash, and it must be fairly waterproof. One of the best I've found for this purpose is the Pentax Zoom 90 WR. I have a number of them set up at food plots, mineral licks, and trails leading into and out of feed fields. The TT-2000 can be tripod- or tree-mounted.

Buckshot 35. This product from Foresite combines a passive infrared sensor, a high-quality Fuji 35mm camera, and a case housing into a compact, waterproof, weather-resistant, airtight one-piece unit that can be hung on a tree or post. It even comes with a heavy-duty camouflage fabric cover. It's computer-controlled, and a two-position switch allows you to select long- or short-range photographs. Because most infrared detectors sense the difference between body temperature and background temperature, a sensitivity adjustment compensates for differences in temperature between summer and winter. The greater the difference, the more sensitive the unit becomes and the farther away it takes a photograph. With no control over the sensitivity of the detector you will get photographs in the winter much farther away than the summer.

A one-minute programmed delay in the Buckshot 35 keeps the camera from taking another photograph until that time expires. This delay, however, can be reprogrammed as desired. The Buckshot 35 is activated or deactivated by bringing a magnet close to the housing. The magnet is also used to check the event counter. The walk-test feature allows you to determine where the subject will be when the photo is taken and also how far away the unit will pick up the subject. I used the Buckshot 35 throughout last year with good results. One especially good feature is the optional "Bear Guard" security harness. A two thousand-pound-test steel chain fitted on a metal harness fits tightly over the unit and secures this valuable investment to a tree so it can't be stolen. Once locked to a tree, it's impossible to remove. I found that in areas with a lot of deer usage I needed to change film every three to four days.

DeerCam NCX10 Scouting Camera. I haven't tested this Non-Typical, Inc. sensor/camera, but have examined it. The passive infrared motion/heat sensor adjusts to a variety of situations. A high/low sensitivity selection allows it to detect game up to sixty feet away. The camera is a high-end Olympus Infinity automatic. Also available is an optional programmer that converts the DeerCam into a full game monitor, allowing you to review the exact date and time animals are in the area before you have the film developed and specify exactly when you want the unit to take photographs. This is an important feature when the camera is used on food plots where deer may spend quite a bit of time. (Without this feature you could end up with lots of photographs of the same deer a few minutes apart and a whole roll of film will be gone in a short amount of time.) Setting the unit to take photographs thirty minutes apart is usually the best choice for food plots. Setting the camera to shoot only during the late evening and night time can also create more selectivity in the animals in case you want to scope out possible big bucks on your food plot without taking too many photos of does and young deer.

Other excellent monitor choices include Stumpcam, Stealthcam, Moultrie, Highlander, and Vision all available from Bass Pro. Cabela's carries Highlander, Moultrie, TrailTimer, Buckshot, Game Country, Game-Vu, and Non-Typical DeerCam.

USING GAME MONITORS

Using game monitors to monitor food plots is as simple as operating a TV or VCR. The more sophisticated units require a bit more learning but also give you more information. Tips for operating monitors:

Make sure the unit is solidly anchored in place. Some units read movement, and if one of these is anchored to a swaying tree or branch, it can give false signals.

Place motion detectors away from high weeds or brush that may blow in the wind and trigger the detector

If you want to check for both turkeys and deer, set the unit about thigh-high. For deer only, set it about chest-high.

Keep records. Although monitors with cameras visually record animal visits, those that monitor and store only event data can also be extremely valuable if you maintain records of times, dates, and numbers of events. Make daily event notations if possible, or print information from the monitor's data base (see the Trail Monitor Log subsection below). Regardless of which type of unit you use, you should be able to determine the most-used food plots and mineral licks, information that can be valuable if you're experimenting with different seeds or materials on different plots and licks.

Game monitors, especially those that have photo capability, are just plain fun. After you go to the expense, time, and trouble to plant food plots or put out mineral supplements, it's a real thrill to open a film package and get a first look at a dandy big-racked buck you never knew existed on your property. In fact, it's downright addictive!

KEEPING A TRAIL MONITOR LOG

Trail monitors will be much more valuable if you keep a daily log of their findings. The more economical models don't record as much information as the more sophisticated ones, but you can still garner a great deal of information over a period of time. The more sophisticated models can even print out charts of daily activity. Over the years I've created a daily log form that has spaces for all of the information I like to keep track of. It fits on a half sheet of paper, so I can get two to a page and it's easy to print up on my computer.

Event Log

Date: ____________________

Occurrences/times: ____

____

____

____

____

Temperature: __________________________ Moon phase: ________________

Weather: _____________________________ Wind: ___________________

Barometric Pressure: ____

If camera-operated

Species: __________________________ Sex: (m) (f)

Age: __________

Comments: ____

____

____

____

Monitoring with Utilization Cages

A very simple and economical method of determining food plot usage is a utilization cage. Deer, turkeys, and rabbits can crop a lot of forage in a short amount of time, beginning as soon as the plants come up. They can also keep a plot eaten down so far that you can't determine how much the plants are growing. A utilization cage is simply a wire basket approximately three feet in diameter and a foot or so high made of welded wire cut and fastened together. Anchor it firmly in place and it will quickly reveal the comparison between grazed and ungrazed food-plot crops.

Food plots, such as those with clover, which may have heavy use can benefit from a utilization cage which allows you to monitor how much is being eaten.

CHAPTER 21

ASSESSING GRASSLANDS

Grasslands are an extremely important part of any wildlife management plan, regardless of whether you're managing for upland game such as quail or pheasants or woodland game such as deer or turkeys. Actually, because of their forage and grain values, grasses are the most important plant group in the world. Grasslands can consist of clearings, which we discussed in Chapter 10, or they can be large, open areas such as pastureland, hay fields, old or idle fields, or fields placed in government-sponsored conservation practices such as CRP (Conservation Reserve Program).

Grasslands are classified in three categories: cool-season, warm-season,

Grasslands are a very important part of wildlife habitat management for many species and in many parts of the country. The health and type of grasslands will determine the amount and diversity of the wildlife.

and native prairies. Cool-season grasses include such species as tall fescue, timothy, Kentucky bluegrass, and orchard grass. Warm-season grasses include the native grasses such as timothy, redtop, switchgrass, big and little bluestem, and other prairie grasses. Warm-season grasses may or may not be native grasses. Native prairies are remnants of native grasslands that have never been plowed or tilled. Native grasses can be planted to make a prairie, but it would never be a native prairie.

The first step in managing grasslands for wildlife is to assess the existing grasslands on your property. A property map or aerial photo is important for this chore so you can superimpose outlines of the fields, mark or identify them, and note acreages. Then you can determine which management practices are needed or desired for the various fields. Use field guides to identify the grasses, weeds, and other plants you find. Since many plants are more productive at certain times of the year, it's a good idea to make your assessment over a year's time. As with many wildlife management practices, grassland management can be costly and time-consuming, especially on a large scale, so you may wish to work on different fields or sections over several years.

Identifying plant species will often help to indicate the health of the grassland and its suitability for wildlife. For instance, lots of broomsedge in an old pasture indicates poor fertility and the need for a managed fertilizer program. And although fescue is a very popular plant for grazing livestock, it's virtually useless as a wildlife grass. On prairie lands the numbers and types of plant species are good indicators of health. Two basic types of prairie species are called decreasers and increasers based on how they respond to management.

Grass Decreasers

Plant species that are preferred by wildlife or livestock for forage are called decreasers. With high palatability, good yields, and high nutrition, they "decrease" in vigor if overgrazed or hayed too short. This is especially so if the overgrazing or haying occurs late in the season. Decreasers include the most desirable of native prairie grasses such as big bluestem (*Andropogon gerardii*); switchgrass (*Panicum virgatum*); little bluestem (*Andropogon scoparius*); Eastern gama grass or corn grass (*Tripsacum dactyloides*); Indian grass (*Sorghastrum nutans*); prairie cord grass or sloughgrass (*Spartina pectinata*); and prairie dropseed (*Sporobolus heterolepis*).

Grass Increasers

The less palatable, less desirable plant species increase in vigor as well as abundance if grasslands are over-hayed or over-grazed, especially late in the season. These include: purpletop or grease grass (*Tridens flavus*); three-awn or wiregrass (*Aristida spp.*); purple lovegrass or ticklegrass (*Eragrostis spectabilis*); winter bentgrass or ticklegrass (*Agrostis hyemalis*); tall dropseed (*Sporobolus asper*); and broomsedge bluestem (*Andropogon virginicus*). With proper management, however, overgrazed or overhayed prairie with increasers as the dominant species can be controlled. This allows the more desirable decreaser plants to replace them. Looking at the species currently growing on your land will give you an idea of past and current land use. Abused grasslands, with grass decreasers more dominant, are in poor health.

Grassland Forbs

Prairie forbs, although considered weeds by many, are actually broadleafed flowering plants. Many are also valuable forage or legumes, readily used by a variety of wildlife. Forbs are also classified as increasers or decreasers and can help indicate the health of your grassland.

Forb Decreasers

The more desirable forb decreasers include lead plant (*Amorpha canescens*); indigo bush, a plant similar to lead plant that grows in wet areas; rattlesnake master (*Eryngium yuccifolium*); pale purple coneflower (*Echinacea pallida*); compass plant (*Silphium laciniatum*); gerardia (*Gerardia tennifolia*); Illinois bundleflower (*Desmanthus illinoensis*); goat's rue (*Tephrosia virginiana*); Virginia lespedeza (*Lespedeza virginica*); ashy sunflower (*Helianthus mollis*); thickspike gayfeather (*Liatris pycnostachya*); rough blazing star (*Liatris aspera*); downy gentian (*Gentiana puberulenta*); sensitive briar, (*Schrankia uncinata*); roundhead lespedeza (*Lespedeza capitata*); and purple and white prairie clover (*Petalostemon purpureum* and *P. candidum*).

Forb Increasers

The less desirable forb increasers include goldenrods (*Solidago spp.*); asters (*Aster spp.*); rosinweed (*Silphium integrifolium*); Indian paintbrush (*Castilleja coccinea*); long-bracted wild indigo (*Baptisia leucophaea*); and white wild indigo (*Baptisia leucantha*).

Fescue, one of the most popular cattle forages, is useless as a wildlife habitat. And, it crowds out other more wildlife-friendly grasses.

Managing Non-Native or Non-Prairie Forage Grasslands

Fescue is considered one of the main reasons for the decline of the bobwhite quail. It grows in a thick, tight mat that quail—as well as young turkey poults—can't negotiate through, yet it doesn't provide overhead cover and protection from predators. And although deer and turkeys may nibble at it, they don't prefer it. It's also extremely invasive and hard to eradicate once started. As many cattle farmers and ranchers have discovered, however, it's easy to grow and sometimes does have a place as forage.

Some other common forage grasses, however, are more "friendly" to wildlife.

Orchard grass. One of the most popular grasses, it grows in bunches like the prairie grasses do. Like fescue, it's primarily a cool-season crop. It provides good, palatable forage for livestock and wildlife as well as cover for ground-nesting birds. If interseeded with ladino clover, orchard grass provides extremely effective grass forage for both wildlife and livestock.

Timothy. This is an excellent perennial cool-season grass that is best suited to cool, humid climates, although we have grown it successfully in Missouri despite the hot summer months. Timothy matures later than orchard grass, is extremely palatable and is a preferred forage for horses. Deer love it. In fact, a large population of deer will soon overgraze a small timothy plot. Timothy is a bunchgrass, so it also is a good choice for ground-nesting birds.

Grasses are either cool-season or warm-season, depending on when they grow. Orchard grass, a "bunch grass" is an example of a good wildlife habitat cool-season grass.

Kobe and Korean lespedeza. Both of these grasses were introduced into the United States from China, Korea, and Japan in the mid 1800s. Kobe is often used for hay by farmers in the South, while Korean lespedeza does best in the upper South and the Midwest. It is a very palatable plant and also provides seeds for ground birds such as quail.

Managing Prairie Grasses

Warm-season native prairie grasses offer great forage for livestock producers and are extremely wildlife-friendly. These grasses, which become fairly tall and grow in bunches, provide nesting, feeding, loafing, and escape habitat for wildlife. They provide good ground cover during heavy snowfall or sleet for birds such as quail, pheasants, and turkeys, and they also offer great bedding spots for deer. Here is a selection of the most popular prairie plant species and their preferred soil types:

Native warm-season grasses are some of the best wildlife grasses. Their bunching growth pattern provides shelter from weather and predators, and allows ground-nesting birds and small game to move freely underneath the upright stems.

INDIAN GRASS *(Sorghastrum nutans)*

Indian grass is a tall sod-forming plant common on deep, moist soils from heavy clay to subirrigated sites. Its sentinel-like bunches wave golden seedheads in September. It is one of the most beautiful of grasses and was named for the Native American Indian. When in bloom, the bright yellow stamens give it a plume-like appearance. It is one of the taller native species, reaching three to seven feet. It has a notched "rifle-sight"-like extension of the sheath portion of the blade that makes it easily identifiable.

SWITCHGRASS *(Panicum virgatum)*

Switchgrass is another tall warm-season grass found in calcareous and wet acid soils. It is a vigorous, sod-forming plant that serves well in grassed waterways and critical area plantings such as highway seedings and dams. It grows in colonies and thrives on moist soils of high fertility.

BIG BLUESTEM *(Andropogon gerardi)*

Considered the king of native grasses, big bluestem is one of the most colorful and robust grasses. After frost, it turns a light reddish color. A sod-forming species, it grows from four to six feet tall on fertile soils or, during wet periods, on shallow gravelly ridges and near limestone ledges. Big bluestem is identified by a three-branched seedhead that gave it the nickname "turkey foot."

Little Bluestem *(Andropogon scoparius)*

The native grass of the state of Nebraska, little bluestem grows on both uplands and lowlands. It grows two to four feet tall and is found in forty-five of the fifty of the United States. In the fall little bluestem turns a beautiful reddish brown.

Sand Bluestem *(Andropogon hallii)*

This tall, warm-season, sod-forming grass is found on dry, sandy lands, primarily in the Nebraska sandhills. A single plant may form a dense colony several feet across and 4 to six feet tall. Sand bluestem helps heal blowouts when wind or water scours the topsoil away in sandy soils. They help control wind and water erosion.

Side-oats Grama *(Bouteloua curtipendula)*

Side-oats grama grows mostly on shallow or limey soils. It is very useful on highway slopes, campgrounds and steep areas. It spreads by short, underground stems and by seed, and it grows only twelve to twenty inches tall.

Eastern Gama Grass or Corn Grass *(Tripsacum dactyloides)*

Preferring medium to deep soils, Eastern gama grass is often observed alongside highway right-of-ways. It's a five- to eight-foot-tall bunchgrass that forms dense clumps one to five feet in diameter. It's the only large prairie grass with a single seed unit arranged in a six- to twelve-inch spike. It's often called corn grass because the seeds are nearly the size of a kernel of corn.

Sand Love Grass *(Eragrostis trichodes)*

Sand love grass has 1½ million seeds per pound and grows well on dry, sandy, upland soils. It is a medium-height bunchgrass that is fine in texture and grows from 2½ to 5 feet tall.

Western Wheatgrass *(Agropyron smithii)*

Western wheatgrass grows well on both lowlands and uplands. It is drought-resistant and winter-hardy. It's a medium-height species that is used in range seedings or western rangelands and for waterways. It will spread into old fields, forming a pure stand. It is a cool-season species growing from one to three feet tall.

BLUE GRAMA *(Bouteloua gracilis)*

This is probably the most widespread of the grama species and grows on a wide range of soils. Blue grama is often referred to as the "Queen of the Plains." It is a short bunchgrass used in range seedings and in such locations as high-use recreation areas and highway medians. A long-lived grass, it has white-purplish flags on each stem and grows from twelve to sixteen inches tall in the North and up to 2½ feet tall in the South.

BUFFALO GRASS *(Buchloë dactyloides)*

Buffalo grass is named after the American bison, and early sod houses were built mainly from it. It's a short (four to six inches), mainly upland plant that grows in patches. It's one of the most interesting grasses because it has male and female plants; the ones that grow flags are males and those without flags are females. Buffalo grass forms dense sod by making runners above ground and by dropping seed, and it's good for erosion control.

REED CANARY GRASS *(Phalaris arundinacea)*

Reed canary grass is often seeded on land too wet for other species because it thrives on wetlands, lowlands, and overflow lands. It is a long-lived plant and grows in clumps as big as three feet across. It's very winter-hardy and is an excellent grass for dams, waterways, and shorelines.

PRAIRIE SAND REED *(Calamovilfa longifolia)*

Sometimes called the "workhorse of the sandhills" because it seems to keep growing in spite of soil and climate conditions, prairie sand reed grows on sandy uplands, though it is more abundant in the sandhills. Conservationists also have a nickname for it—"sandbinder."

PRAIRIE CORD GRASS OR SLOUGHGRASS *(Spartina pectinata)*

A five- to twelve-foot sod-forming warm-season grass, this plant prefers bottomland soils and wet draws and often forms pure stands. The veins of the leaf are prominent, and the upper surface and margin of the leaves are rough to the touch, which has given it the name of "rippey grass" or "ripgut." In the early stages it is used for hay, and it's also great for covering waterfowl blinds.

Once you determine the health and species of the grasslands on your property, you can begin a management plan. First set goals. Will you be managing specifically for wildlife? Will you be managing for both wildlife and farming or ranching practices, such as grazing? If you're managing for wildlife or wildlife in conjunction with a grazing or haying operation, the next step is to determine what type of wildlife you wish to manage for. Although you may choose one species, often other species will also benefit from the management plan. The following chapters detail general management as well as reestablishment of grasses.

CHAPTER 22

General Grassland Management

Grasslands require management to be productive for livestock, haying, and wildlife. If left idle, ground litter builds up, and this reduces the amount of palatable forage and restricts the movement of wildlife, including quail, pheasants, turkey poults, and rabbits. Increased ground litter buildup also increases rodent populations. Five basic practices are used in grassland management: fertilizing, overseeding with legumes, grazing, haying, and prescribed burning.

The first step in grassland management is taking a soil test and bringing the grass or area to be planted up to soil test as well as adjust the pH with lime.

Fertilizing

Before you do anything else, take a soil test. The results will include recommendations for adding fertilizer and, usually, lime.

Soil Sample

Soil often varies quite a bit within the same field. A good soil sample is a composite of samples taken up to seven inches deep in at least ten locations of the same type of soil in a field. Test the soil as early in the spring or as late in winter as you can. Remove surface residue from the site before taking the test and use a spade to scrape away vegetation and roots. Soil tests may be taken with an auger or soil tube, or you can use a spade to cut a seven-inch-deep trench and then cut a one-inch slice from the side of the trench.

The individual samples should be collected in a clean bucket or other container (don't use a galvanized pail—the zinc may contaminate samples). Mix the samples thoroughly, and then place about a pint of the mixture into a soil sample box, bag, or clean container such as a plastic food container. If the soil is wet, spread it out and allow it to dry first, but do not apply artificial heat to aid in drying.

Label the sample with the field number or name and mark the field number or name on an aerial photo of your property. It's important to include the following information with your samples and to keep a copy for yourself: acres represented; last crop grown; dates of liming and the amount applied per acre; dates of rock phosphate use and amount applied per acre; irrigation and land leveling practices or plans; kind and amount of fertilizer applied the previous year; soil characteristics such as bottomland, upland, rockiness, and topography; cropping system planned for the next four years; and yield goals.

Take the soil samples to your local county extension center or a private laboratory. Drying, processing, testing and interpretation your samples usually takes about two weeks. With the soil sample results, you're ready for the first step in management—bringing soil fertilization up to soil test recommendations and adjusting the pH. It's important to understand how the different elements affect grassland productivity.

Understanding and Applying Fertilizers

- Fertile soil is necessary to grow plants, including food plots. The fertility of the soil is dependent on several factors:
- An adequate supply of plant-food elements.
- Sufficient moisture to supply the plant foods to the plant roots.
- Enough warmth to encourage plant growth.
- Air to supply oxygen.

The organic makeup of the soil is important. Friable or loose soils with organic matter, including soil bacteria and necessary minerals, grow the best plants. Heavy clay-type soils are not as productive because the plant foods and moisture can't get to the roots as well and less oxygen is available. Sandy, dry soils don't hold moisture and also allow plant foods to escape from roots. Different types of soils hold different amounts of nutrient elements or cations: [calcium, magnesium, potassium, and the non-nutrient cation hydrogen (neutralizable acidity)]. The capacity of soils to hold these cations is called Cation Exchange Capacity, and it depends on the kind and amount of clay and organic matter in the soil.

Positively charged ions, known as cations, are held by negative charges of the soil colloids (clay and organic matter). Plant roots exchange hydrogen ions (acidity) for calcium, magnesium, and potassium ions. The resulting effect on the soil is depletion of the nutrient ions and increasing acidity. Soil treatment becomes a matter of replacing neutralizable acidity with a satisfactory balance of calcium, magnesium, and potassium.

Desirable values of exchangeable cations for Balanced Soil Saturation are seventy-five percent calcium, ten percent magnesium and 1.7 to 5.3 percent potassium for heavy clays to sandy soils respectively. Saturating of calcium may vary from fifty to seventy-five percent and magnesium from six to thirty-five percent providing that the sum has a minimum value of eighty-five percent saturation.

Providing Soil Nutrients

As explained above, plants need nutrients, including magnesium, phosphorus, potassium and nitrogen.

Magnesium. Magnesium (Mg) is found in limestone. The amount of magnesium carbonate per ton of ground limestone varies.

Phosphorus. Phosphorus (P) has many important functions in plants, the primary one being the storage and transfer of energy through the plant. Adenosine diphosphate (ADP) and adenosine triphosphate (ATP) are high-energy phosphate compounds that control most processes in plants including photosynthesis, respiration, protein and nucleic acid synthesis, and nutrient transport through the cell walls. Phosphorus is essential for seed production, promotes increased root growth, produces healthy growth, and encourages good fruit development. Rock phosphate is the most common source of phosphorus.

Potassium. The most common source of potassium (K) is muriate of potash, derived from large deposits of potassium chloride salts found in the southwestern deserts of the United States. Potassium is needed for the manufacture of carbohydrates—sugars and starches. Potassium also increases resistance to disease and produces strong plant cell walls and stems.

Nitrogen. Nitrogen (N) is necessary to convert the sun's light into energy through photosynthesis. Plants also use it to form amino acids, the building blocks of protein. Protein is needed by all types of wildlife and the availability of plant-usable nitrogen often determines the quantity and quality of forage available. Plants with dark green leaves have a high or proper amount of nitrogen availability. Slow-growing, stunted plants or those with yellowing leaves show a nitrogen deficiency.

Each plant has different nitrogen requirements. Corn, for instance, has one of the highest nitrogen requirements, needing as much as 215 pounds of nitrogen per acre to produce high yields. Legumes, such as clovers, alfalfas, and soybeans, "fix" or add nitrogen to the soil. Rhizobium bacteria form nodules on the roots of the plants and then take nitrogen from the atmosphere and make it available to the plant. Although these plants add some nitrogen to the soil, they often use most of it as well. If they are planted with other plants, you may need to add some nitrogen to the soil. Applying nitrogen early in the season, however, allows grasses and weeds to get a jump on the legumes. Apply the nitrogen in the late spring or, better yet, late summer.

Overseeding with Legumes

A simple way to add value to grasslands is to overseed them with legumes. The legumes, including clover and lespedeza, take nitrogen from the air and add it to the

One simple method of enhancing existing grasslands is overseeding them with legumes. The legumes add vital nitrogen to the soil and provide excellent wildlife food.

soil, which makes nitrogen available to the grasses and improves forage quality. The legumes also offer food for wildlife. Almost all animals, including deer, turkeys, and rabbits, love the clovers, and quail and turkeys relish the seed from the lespedezas. We've done this on our farm almost from the beginning back in the early '70s.

Legumes will do poorly if seeded into fields with a low pH—one less than 5.5. Because lime moves slowly through the soil, apply lime (based on the results of your soil test) well in advance of seeding. Phosphorus is especially important for legume establishment. Unless the stand is high in phosphorus, apply some at seeding time. Potassium is not as important at seeding time, but longevity of the stand can be increased when potassium is used as a top dressing.

Establishing legumes in thick stands of fescue can be difficult. Legumes establish best in bunchgrasses such as orchard grass and timothy. They also establish fairly easily in bluegrass because it is short.

Overseeding an established pasture or hayland can be fairly simple, but the results will vary according to the type of existing vegetation and the state it is in. Three methods can be used to reestablish legumes in grass sods with a minimum of production. In all three, the idea is to control the grass so the young legume plants can compete. As with all seedings, it is important to supply adequate moisture, light, and fertility. The rate of seeding depends on the method of sowing.

Winter- or Frost-Seeding

This is done by spreading seed either on snow or on frozen ground. As the ground thaws the seed is percolated into the ground and lightly covered. Before you spread seed, condition the grass by overgrazing the area in fall and early winter and apply the amounts of lime, potassium, and phosphorus specified by the soil test during the same period. Do not apply nitrogen, which will stimulate the older grasses and weeds and provide competition for the young legume plants.

Inoculate the legume seeds if they're not preinoculated and broadcast them early enough in late winter or early spring that freezing and thawing will cover them. Seeding in February usually results in better success than seeding in March or April. Although new seedlings can occasionally be injured by a late freeze, earlier is better than later so that the legumes get a good start before the grass starts to compete with them.

Late Fall/Early Winter Tilling

In late fall or early winter, till the sod so that about fifty percent of it is disturbed. An offset disk works well for this. Broadcast or drill the legume seeds directly into the partially opened sod in late winter. Apply fertilizers other than nitrogen. Keep the early grass growth down by clipping or grazing.

Using a Chemical Grass Growth Retardant

Use a chemical such as Paraquat to retard grass growth until the legumes get a start. A rate of one to two pints per acre is considered adequate. Using too much will retard the growth of the grass too much, resulting in excessive weed growth. Seed during the growing season, from early spring through late summer, with no-till grass drills such as the Great Plains models.

No matter which method you use, always buy certified seed with proven performance and determined purity. Always inoculate unless you use preinoculated seeds. Mix the inoculant with a liquid—soft drinks work well because their stickiness helps the inoculant to adhere to the seed and moisten it.

New seedlings must be properly managed during the year they're established. Graze the new seedlings in the spring until the tops of the young legumes are being eaten or top-clip with a mower. Then allow the pasture to rest for four to six weeks. After that, practice rotational grazing (see next section) until about the first of September. Between September and the start of cold weather, don't graze or clip the stand. The key is not to overgraze during the first year of establishment.

Information about Specific Legumes

Clovers

Ladino clover is one of the easiest legumes to establish in most grass sods, especially in thick, heavy stands of grass. It does not, however, do well during the dry summer months and is less drought-tolerant than some other legumes.

Red clover is also easy to establish and grows on a wide range of soils. It needs to be reestablished every two to three years.

Alsike clover is fairly short-lived on droughty upland soil, but it is a perennial on wet lowland soil. It does not regrow when cut in spring except in very wet conditions. Its best use is as a specialty clover in poorly drained areas. Some people use two to three pounds of alsike clover along with half a pound of ladino clover per acre.

Alfalfa

Alfalfa requires more management than the other legumes, including more lime, more fertilizer, and spraying for alfalfa weevil, and it needs soil with a good internal drainage. It is fairly drought-resistant and is a favorite of wildlife, especially deer and turkeys.

Lespedeza

Kobe or Korean lespedezas grow on a wide range of soil types. They are fairly easy to establish and maintain.

Grazing

Most folks don't think of grazing as a method of wildlife improvement, but with many grasslands, following proper grazing practices can definitely be a benefit. Grazing is either continuous or rotational.

Most landowners don't think of grazing as a wildlife management tool, but done properly it can be extremely effective. Intensive grazing, using rotational paddocks, maintains grassland vigor.

Continuous Grazing

Continuous grazing offers little benefit to wildlife, to the health of the grassland and, in the long run, to the livestock, and it has several drawbacks. First, livestock are very selective about what they eat. They tend to overgraze the plants they like—the most palatable grasses—and undergraze the less desirable plants. This allows the undesirable plants to overtake the more desirable plants. After years of overgrazing, pastures change from grasses to brush and weeds. Continuous grazing also creates areas in the pasture where livestock congregate, often denuding vegetation, creating erosion problems, and trampling wildlife nests. Over the years, continuously grazed grasses have less vigor, become less productive, and provide less wildlife food and cover.

Rotational Grazing

Rotational grazing can be done in several ways. Old-timers simply switched livestock between two pastures, allowing one pasture to rest while the other was grazed. If they had the land, livestock was switched among several pastures in rotation. A much better plan used by livestock producers these days is intensive grazing. We were one of the first in our area to establish this practice on our farm, and the results have been very impressive. Intensive grazing allows much better utilization of the land, provides better wildlife habitat, generates better health for the grassland, and actually increases livestock gains (gains in the weight of the livestock). The method is also extremely simple. Divide a pasture into small paddocks, the size determined by the number, size, and species of the livestock herd. Each paddock may be as small as two to three acres. Allow livestock to graze the area until they have eaten the grass down to a predetermined height. (The small amount of space forces them to eat all species.) Then turn the animals into the next paddock, the next, the next and so forth until they rotate back to the first paddock. Our cattle are rotated once a week, usually Saturday morning, and it takes about an hour. They are more than ready for the change and wait at the gate for the next move. This method creates more manageable livestock, as they are more dependent on you and used to having you work with them.

Although permanent fencing can be used to create the paddocks, portable electric fencing using plastic polywire and push-in electric fence posts makes defining them easy. In one forty-acre pasture we keep two paddocks fenced at all times. We simply leapfrog the last polywire as we move the cattle in a wagon-wheel-type rotation around the pasture and the water source. A pond with a below-dam frost-free water tank is located in the center. The temporary fencing also allows us to take the livestock off the field entirely for a season and cut the field for hay.

Another type of rotation is moving livestock between cool-season and warm-season forages, and this offers many wildlife benefits as long as you don't allow the animals to overgraze. Cool-season grasses can be grazed two to 4 inches tall. Warm-season grasses shouldn't be grazed shorter than eight inches and they shouldn't be grazed until midsummer.

Haying

One of the most important grassland management practices is haying. Grasslands that are not occasionally cut for hay eventually sprout into saplings, shrubs, and weeds. Haying must be done properly, however, in order to produce the best and most forage, maintain a healthy grassland, and provide habitat for wildlife, including protecting nesting cover and nests.

Haying is an important factor in many grasslands. It's important not to hay too early or ground nests as well as quail, pheasant, and turkey hens may be destroyed. Usually the last of June is considered safe for most areas.

Cut for hay at the proper time. Cutting too early reduces forage production and may destroy wildlife nests. Cutting too late doesn't allow the grasses to replenish their root reserves before the winter months and weakens the stand.

Raise the cutter height to four inches when cutting cool season grasses to reduce the chance of disturbing ground-nesting wildlife.

Don't cut warm-season grasses for hay in the fall.

Cut at the proper height. If you cut too high, there's little production. If you cut too low, regrowth is slower in the grass stem.

Be aware that haying has a dramatic effect on wildlife survival. With the cover gone, all types of wildlife are more vulnerable to predators.

Performing Prescribed Burns

Native prairie grasses and native prairies benefit greatly from properly done prescribed burns. The fire controls the ground litter, allows for better release of nutrients, controls many unwanted plants, stimulates seed production in many desirable plants, and creates a more diverse plant community. Prescribed burning is covered in the next section.

Combining Grazing and Burning

In recent years managers of prairies have done a number of studies on the combined use of fire and grazing. They've determined that prairie wildlife species, such as prairie chickens and quail, as well as livestock grown on prairies, do best with a checkerboard-type management system in which burning and grazing are done in rotation.

CHAPTER 23

GRASSLANDS ESTABLISHMENT

In many instances you may need to establish new grassland areas. This may be because your pastures are old and deteriorated or because you wish to create warm-season grasslands. The latter are especially beneficial for wildlife.

ESTABLISHING NATIVE PRAIRIE GRASSES

There are several ways so establish warm-season native prairie grasses. Regardless of the method used, it takes patience and time to achieve success. It takes two to three or more years for most seedlings to flower and at least three years for prairie plants to overcome the initial weeds and grasses.

SEED FOR WARM-SEASON GRASSES

Establishing native warm-season grasses is not easy and it's also fairly expensive. The results, however, are more than worth it when managing grasslands habitat for wildlife.

Native warm-season grass (NWSG) seed is expensive, and it takes quite a bit of effort to establish. For this reason, buy seed only from a reputable dealer. Purchase only Pure Live Seed (PLS) and only the amount you need. PLS means the amount of seed minus the trash. To find the

PLS of a batch of seed, multiply the purity percent (P) by the germination percent (G), and then add the firm seed (F). Then multiply the PLS percent by the weight of the seed to give the pounds of PLS. The figures for the percentages are usually found on the bag or provided by the seed dealer.

Native warm-season grasses can be planted in pure stands or mixtures. Mixtures are the best choice for wildlife plantings. Good combinations include big bluestem and Indian grass, or big bluestem, Indian grass, little bluestem, and side-oats grama. Switchgrass is usually not included in mixtures but is seeded by itself into low or wet spots. Seeding rates for mixtures run from 1 to six pounds per acre in these types of plantings. A seed rate chart is included at the end of this chapter.

NWSG may be seeded in the spring or fall. Spring seeding should be done from April through June. Fall planting can be after November 1st when the soil has cooled down enough that the seeds won't sprout until the next spring. In this case, seeding rates should be increased twenty-five to fifty percent.

Additions to grass plantings:

- *Legumes.* Legumes added to the NWSG stand can increase yields for livestock or hay and increase usage by wildlife. Broadcast Illinois bundleflower, a native legume, at a rate of one-quarter to one-half pound per acre with the grass seed. Overseed or broadcast Kobe or Korean lespedeza at five pounds per acre during January or February just prior to the third growing season.
- *Native forbs.* Native forbs can also add to the overall attractiveness of the seeding for wildlife, and many of the forbs are extremely beautiful as well. Forbs can be added to the seed at time of planting, or they can be overseeded in January or February.

Soil Testing and Fertilizing

As with all plantings, before you plant, take a soil test and apply the needed lime, phosphorus, and potassium prior to seeding. Do not, however, apply nitrogen. Most grass varieties require at least a pH of 6.0.

Ground Preparation

Properly preparing the ground is extremely important for successful establishment of NWSGs. The seedbed should be extremely smooth and firm.

Ideally it should be cultipacked or rolled. On lands that can be plowed, plow in late fall, then disk to kill weeds until planting. On shallow soils, disk instead of plowing. One of the traditional methods, where applicable, is to grow a cultivated crop two years prior to converting to NWSG.

Erosion control. Because of the time it takes for vegetation to emerge, it's important to control erosion. This means providing some form of vegetative cover until the warm-season grasses can take hold. Four basic methods can be used:

- *Seeding into milo or soybean stubble in the late spring.* After harvest in the fall, leave the stubble standing and then mow to four to six inches prior to planting.
- *Sow a cover crop of oats in late summer or early fall,* and then plant NWSG into the standing mulch the next spring.
- *Plant two pounds of timothy, redtop or orchard grass* early in the spring so these cool-season grasses will sprout before the warm-season ones do.
- *Kill the vegetation with herbicides and then sow in the mulch.*

Planting

Plant seeds using one of two methods:

No-till drill using a rangeland drill, such as the Great Plains models. Local Soil and Water Conservation Districts have grass drills for rent. Some state conservation agencies also rent them as well, and you may be able to find a local contract operators to do the job for you.

Broadcast seed using a broadcaster on a tractor, truck, or ATV. Because the seed is fluffy, you'll need a broadcaster with a bottom mixer blade. Mixing the seed with phosphorus and potassium fertilizer or oats helps distribute the seed more evenly and prevents it from packing into the bottom of the spreader. Cultipack or roll the area before and after seeding.

The number one mistake people make in seeding NWSG is planting it too deep. NWSG seeds should be covered with no more than a quarter inch of soil, and it doesn't hurt for some seeds to be exposed. We've experienced sparse germination after drilling too deep. Our most successful fields have been broadcast on a prepared seedbed which was then rolled with a field roller. Don't harrow or disk after seeding.

Here are some spring and fall seeding practices.

To prepare the ground for native warm-season grasses, kill back existing vegetation, which is quite often fescue. Fescue should be mowed, burned, and then an herbicide should be applied to the soil.

Spring

- *Mow or burn in the winter to encourage new growth.* Once plants have reached the boot or bud to early seedhead stage in spring, apply two quarts of glyphosate (Round-up) in twenty gallons of water per acre. Add six ounces of nonionic surfactant per twenty gallons of water per acre. Use 3½ pounds of ammonium sulfate per twenty gallons of solution. All chemicals are mixed together and applied as needed. For non-CRP acreage, use twelve ounces imazapic (Plateau) plus two pints of MSO per acre after green-up of fescue and before boot stage.
- *If you're planting on a prepared seedbed, wait one to two weeks, then disc.* For non-till seeding you can drill immediately.
- *If you can't plant the NWSG/forbs in time, plant a summer cover crop.* Apply one quart per acre of Round-up in the fall to clean residual fescue. Plant NWSG seeds during the dormant period the next spring, leaving the cover crop as mulch.
- *We've discovered the best tactic (instead of just Round-up) is to use*

a combination of Plateau and Round-up. This provides better weed control than either herbicide by itself. The mixture also does a better job of killing fescue. Tank mix 1 to 1½ quarts Round-up plus four to eight ounces Plateau. (On CRP land you must use only four ounces imazapic—which is the same as Plateau.) Add seventeen pounds of ammonium sulfate per one hundred gallons water and apply at twenty gallons of water per acre.

- *Another method is to apply the Round-up plus ammonium sulfate.* Wait one to two weeks and burn the killed residue. Then plant NWSG. Apply eight ounces of Plateau (four ounces on CRP land) a few days before or just after planting. Our most successful practice has been to apply the combination mix of herbicides, wait two weeks, and then thoroughly disk, cultipack or roll, seed, and roll again.
- *It's least costly to use a cropping method, or planting a row crop such as corn or soybeans and then harvesting it, on lands that are tillable.* The crop income can help offset the establishment cost. Apply one quart per acre of Round-up in the fall or two quarts per acre in the spring. Then plant a summer crop such as soybeans or milo. Use weed control herbicides, but not imazaquin herbicides such as Squadron, Triscept, Steel, or Scepter, which may have a carryover that can affect the establishment of NWSG.

Fall

Both of the following fall-sown methods reduce adverse effects on existing native grasses such as side-oats grama, Eastern gamagrass and susceptible native forbs such as compass plant, ashy sunflower, and rosinweed.

- *Simply use repeated prescribed burns and/or selective herbicides to kill out unwanted grasses and weeds, allowing native grasses to re-establish naturally.* Imazapic (Plateau) is the most effective herbicide for this practice. Apply four ounces of Plateau plus two pints of methylated seed oil (MSO) per acre. This practice does require several years of repeated burn and herbicide application, but is a common method used on CRP lands or areas that may have native prairie seeds lying dormant. This method may not completely control tall fescue.
- *A better method is to use Plateau but also seed the area.* In this

Warm-season grasses can be drilled or broadcast. If the latter, a firm, smooth seedbed must be established first.

case, mow, hay, or burn the area in July or August to reduce residue and ensure maximum exposure of new growth to the herbicide. Then apply twelve ounces of Plateau plus two pints of methylated seed oil (MSO) per acre. Plant the NWSG and forbs during the dormant period or the following spring.

Native grasses can also be broadcast or drilled in spring or fall. One of the simplest methods involves no-till fall broadcast seeding of native prairie grasses. The existing vegetation must first be killed using an herbicide. Glyphosate (Round-up) is most commonly used and should be applied three times before planting—in spring, summer, and early fall for a fall sowing. Brushy weeds such as blackberry, sumac, sassafras, and trumpet vine can be a problem if they're not killed before you sow prairie seeds. A stronger herbicide like Garlon or Pathway may be needed.

After existing vegetation has been thoroughly killed, don't till, disk, or plow the area, because disturbing the soil tends to bring up extra weed seeds. Cut dead vegetation to a few inches high using a rotary mower. Having some dead vegetation present will help hold the seeds in place and prevent erosion. Seed directly on the prepared area using a broadcast seeder. Fall sowing should be between November and January. Sow ten pounds of cleaned mixed seed per acre (fifteen pounds if seed is chaffy). Mix the seed with a larger volume of slightly moist sand, sawdust, or similar material for improved seed distribution. Forbs can be mixed in with the seed or broadcast during the winter dormant season.

Managing Established Grasses

It is important to keep weed growth under control during early stand development. Don't let weed grow more than eighteen inches tall, and clip them to four to six inches with the first mowing and six to eight inches with the second mowing. After August, clip the weeds high enough to avoid cutting the new grasses or stop mowing entirely.

You can also allow livestock to "flash-graze" the area. To do this, allow cattle to graze just enough to allow them to eat the weeds, but not the developing grasses. Don't allow them to stay on a particular area more than two days each week. When the weed crop is three inches tall, allow grazing again. Stop all grazing by the first of August.

Cool-season grasses mixed with legumes also provide excellent wildlife habitat. The same basic steps are required for establishment—soil-test, fertilizing, liming, killing back fescue, tilling, and seeding.

You may also use a light cutting or grazing as management the second year, but not after July 15. Don't mow or hay after August 1. Overseed with Korean or Kobe lespedeza following the second growing season. After the third year, normal usage can begin. Prescribed burning can be used as a management tool after the fourth year.

Establishing Cool Season Grasses

Fall is a good time to establish cool-season grasses such as orchard grass and timothy as well as mixtures containing legumes such as clover or alfalfa.

Have a soil test done and then apply lime to adjust the pH.

Plow deep soils or disk shallow soils in early summer. Leave the soil bare or fallow until seeding time. The idea is to have the soil loose and rough, exposing it to sunlight and rain to let the weed seeds germinate and then to destroy the weeds with regular, light, minimal cultivation.

If you're renovating cool-season grasses into heavy fescue, kill the fescue using the herbicides mixed as described in the NWSG section above.

Work the seedbed until it is firm and fine-textured using a disk, harrow, and cultipacker or field roller about the first to the second week of August. The latter will also help flatten, smooth and bury rocks.

Sow the seed and fertilizer as soon as possible after the middle of August or a month or so before the first frost. They may be broadcast with the proper amount of fertilizer or they may be seeded with a drill. When you broadcast seed, use fifteen to twenty percent more seed than when you use a drill. If you're broadcasting, cultipack or roll the seed in place. Cover seeds with about a quarter inch of soil. Here are some good wildlife mixtures:

- *Eight pounds orchard grass and fifteen pounds alfalfa per acre.* This is also a good seeding mix for hay.
- *Eight pounds orchard grass, fifteen pounds alfalfa and eight pounds red clover per acre.*
- *Eight pounds alfalfa, two pounds ladino clover, two pounds timothy and four pounds orchard grass per acre.* This is an area that can also be used for pasture.

Do not graze or harvest newly seeded areas until eight months or more after seeding to allow deep root set and to avoid stunting of plants. Top-dress fields with fertilizer according to soil test.

SEEDS	SEEDING RATE (pounds live seed/acre)
Alfalfa	12-15
Birdsfoot trefoil	4-8
Caucasian Bluestem	3-4
Clover, alsike	4-6
Clover, crimson	20-25
Clover, ladino	1-3
Clover, red	8-12
Eastern gamagrass	10 (drill)
Indiangrass	6-8
Lespedeza, Korean	10-15
Orchardgrass	10-15
Orchardgrass+Alfalfa	6+10
Orchardgrass+Ladino clover	6+1
Orchardgrass+Lespedeza	6+15
Orchardgrass+Lespedeza+Ladino clover	5+15+1/2
Orchardgrass+Red clover	6+8
Switchgrass	6-8
Timothy	6-8
Timothy+Birdsfoot trefoil+Kentucky bluegrass	1+5+2
Timothy+Red clover	2-4+8
Triticale	70-100
Wheat	100-150

CHAPTER 24

ADVANTAGES OF PRESCRIBED BURNS

For more than fifty years, Smokey the Bear has warned that "only you can prevent forest fires." While his message is intended to save forests, the resulting suppression of fires has instead increased the chances of catastrophic wildfires across much of the United States. It has also resulted in the reduction of wildlife populations in some areas.

WILDFIRE

After a wildfire, a forest often looks like a barren wasteland. Among pillars of charcoal barely recognizable as trees, the only things that seem to remain are ash, dirt, and rocks.

"But this picture is not grim," said Dan Dey, a research forester stationed at U.S. Forest Service's North Central Research Station on the Missouri University (MU) campus. According to Dey, who specializes in disturbance ecology, wildfires can be beneficial to a forest system. He explained that when a fire is introduced into a system, some individuals may suffer, but the population is usually better off overall. He listed the following as some of the benefits fire can bring to a forest:

Fire can be one of the worst problems with wildlife habitat management or, if used properly, it can be one of the best. Fire creates diversity, reduces insect infestation, and creates new, more succulent plants. COURTESY ATV COUNTRY, INC.

Increased biodiversity. After a fire, different habitats develop in forest openings. This creates a mosaic of new habitats that encourages new species to enter the area.

Increased productive growth. Because not all plants survive, those that thrive without as much competition grow faster than they could before. The new plants are also more succulent, more palatable, and closer to the ground, making them better food sources.

Reduced probability of disease or insect infestation. Fire can improve the health of a forest by removing trees weakened by insects or disease and decrease the chance of future infestations and outbreaks.

Dey added that whether these and other benefits are realized depends on several factors, including a fire's intensity, frequency of fires, the season in which the fire occurs, and the extent of burning.

According to Josh Millspaugh, MU assistant professor of wildlife conservation, most wildlife species adapt to survive wildfires, and in some cases they benefit because of them.

For six years, Millspaugh studied how elk in the Black Hills of South Dakota reacted after wildfire burned forty percent of the area's forested habitat in 1988 and 1990. He found that although the elk had to deal with many new stresses, such as the loss of cover, they adapted and thrived in the new environment, doubling their population.

Although not all wildlife are as adaptable as elk, Millspaugh lists the following ways fires can potentially benefit some wildlife species:

They improve the quality and quantity of food for large herbivores. Fire removes old woody vegetation that is low in nutrition and allows new plants to grow that are better food sources for species such as elk, deer, and moose.

They improve habitat for some bird species. Birds that nest in cavities, for instance, benefit from fire because after the fire there are more dead trees in which they can build a home.

Millspaugh said that many wildlife species have evolved with fire and are able to adapt after a disturbance.

Wildfires, however, are usually not the best way to improve wildlife habitat. They can be costly and dangerous. Extremely intense and long-lived fires can also reduce the improvements that are usually found in the soil after a fire.

Although a number of variables—climate, topography, vegetation, and soil type—determine how soil is affected, wildfires under natural conditions typically consume dead plant matter near the ground and release the nutrients contained within them. These nutrients are then available to the surviving plants, allowing them to thrive in a less competitive environment.

Extremely intense, all-consuming fires, on the other hand, can turn entire forests to ash, leaving very little material to hold nutrients such as calcium, magnesium, nitrogen, and potassium in place. "In some instances, wildfires can even bake the soil and make it water-repellent, preventing any nutrients from being absorbed," said David Hammer, professor of soil and atmospheric sciences at MU. Hammer added that while these nutrient losses may last only a few months, the effects of their losses may be felt for many years, altering the forest's ability to heal long after reseeding crews have spread new seed.

Richard Guyette, an MU forestry professor, contends that fire, if used carefully and correctly, can be beneficial. Guyette, whose specialty is dendrochronology, the study of tree rings, has found that fire frequency and human population density are closely related. His research has shown that this human-fire relationship can be divided into four succession stages:

1. *Ignition-limited.* Human population density is low, resulting in fewer fires set by humans.

2. *Fuel-limited.* Fires are independent of population and limited fuel production (i.e. when fires burn themselves out because they run out of fuel).

3. *Fuel-fragmentation.* Population increases leads to division of fuel, limiting fires (i.e. when land that was previously forested is cleared, leaving islands of forest instead of a contiguous forested plot, and fires are then limited by the size of the forest).

4. *Culture-limited.* Attitudes toward fire lead to suppression; fuel builds up over time.

Guyette says the United States is in the culture-limited stage; people listen to Smokey's messages and and are careful not to start forest fires. He warns that the longer this occurs, the greater the chance of disasters from the buildup of fuel. Guyette encourages returning to the fuel-limited and fuel-fragmentation stages, especially in urban areas adjacent to wildlands. More frequent low-intensity fires would reduce fuel levels, he says, helping to prevent destructive wildfires.

Prescribed Burns

Prescribed burns, sometimes called controlled burns, can be used to manage forests, grasslands, and even marshlands. Done correctly and at the right time, prescribed fires can be extremely beneficial.

They alleviate the build-up of fuel. This can really prevent serious, out-of-control wildfires, especially in forestlands. In grasslands, fires can eliminate vegetation that has become extremely thick, leaving fewer protection and movement areas for ground-nesting birds such as quail, pheasants, and turkey poults.

Prescribed burns can be used to suppress certain species that are less fire-tolerant, yet improve conditions for other species. A good example in woodland areas is cedars, which can take over a hardwood forestland. A prescribed burn is a common practice for cedar reduction. In grasslands fescue can be controlled or even killed, allowing native warm-season grasses to reestablish.

Succulent forbs quickly come back after ground litter and some low-quality shrubs are killed in woodland areas. These forbs are relished by all types of wildlife.

Some animals, such as ruffed grouse, elk, deer, moose, bear, and prairie chickens, are fire-dependent. In other words, over the past half million years these animals have become dependent on the quality of habitat produced by occasional fires.

The amount of cover available for predators is reduced.

The recycling of minerals and nutrients is accelerated.

Sucker growth is stimulated and conifer growth is reduced in aspen forestland.

Fire has always been an important element in the health of prairies, plains, and forests, but especially so in native prairies. In fact, the spread of the eastern forestlands into the western plains has been a direct result of fire suppression.

Prescribed burning controls woody plants and herbaceous weeds, stimulates desirable plants, improves poor grazing distribution, reduces wildfire hazards, increases livestock gains, and improves wildlife habitat.

Protecting Your Home from Wildfire

As urban development continues to blur the line between city streets and country roads, homeowners face unique challenges when protecting their property from wildfire.

According to Bruce Cutter, another MU forestry professor, living along what is known as the "wildland/urban interface" can be dangerous if people aren't careful. Cutter, who is also a firefighter with more than twenty years of experience, said that many homeowners in these areas unfortunately expect urban response times in a rural environment.

Cutter cited access as one of the greatest problems with homes built in the wildland/urban interface. Many homes are located along narrow, winding roads and have narrow driveways and narrow or covered bridges that make getting large trucks to a fire difficult. To protect homes and property, Cutter offered the following suggestions for landowners in these areas:

Create a low-fuel zone between property and adjacent woodlands. This can be accomplished by keeping your lawn well mowed, removing trees that are next to or hanging over the house, and moving firewood piles away from the house.

Implement a prescribed burning plan. By burning periodically, fuel levels are kept low and the chance of a catastrophic fire is diminished. Consult your local Natural Resources Conservation Service office or state forester for more information.

Before burning, contact the local fire department. If the fire department knows you are burning a brush pile, it can be ready to respond if needed and they'll be aware of the fire if your neighbors call to report it.

Visit a fire safety web site. Firewise (www.firewise.org) provides reliable fire safety information to homeowners in regard to everything from fire hazard assessment to landscaping.

CHAPTER 25

IMPLEMENTING PRESCRIBED BURNS

Prescribed burns are also called "controlled burns," but unfortunately, sometimes such burns can't be controlled. Several years ago, a burn started by experts in our part of the country got out of control and a couple of automobiles owned by the burn controllers were lost. Several wildfires in the West have also been attributed to the loss of control of a prescribed fire. Proper timing, having the right tools, and understanding how to plan and conduct a prescribed burn are extremely important. Many Soil and Water Conservation District (SWCD) offices, as well as some state agencies, hold "burn schools." If you're interested in implementing prescribed burns as a wildlife management tool on your property, check with local forester or SWCD offices and attend one of these schools.

TIMING AND FREQUENCY OF BURNS

The timing of the burn is extremely important. Plants are affected differently depending on the time of year the burn takes place. How often you burn also affects plant growth. Before you burn, determine what your goals are—eradicating fescue, increasing or managing native-warm-season grasses, etc.

BURNS TO ELIMINATE FESCUE

To kill or eradicate fescue, burn in the fall or late winter. Burning in the fall reduces carbohydrate reserves throughout the winter and causes slow regrowth in the spring. It can, however, cause erosion problems. After five to six inches of green growth in the spring, spray it with Round-up and Plateau to kill the

A prescribed fire can be used for several purposes, such as killing back fescue, eradicating brushy growth in old fields, or reducing cedars in timberlands. COURTESY ATV COUNTRY, INC.

fescue. Then interseed the desired species into the dead tall fescue stand. If the tall fescue returns, burn the field in April of the following year to reduce competition with warm-season grasses and forbs.

Burns to Improve Stands of Cool-Season Grasses

Stands of cool-season grasses—such as orchard grass—can often be improved with a burn just after the grasses break dormancy, which is usually in February. This increases plant density and improves forage quality. Burning at this time also encourages forbs, legumes, and annual warm-season grasses to grow.

Burns to Improve Stands of Native Warm-Season Grasses

Big bluestem, Indian grass, Eastern gama grass, switchgrass and the other native warm-season grasses respond the most positively to burning, which helps to maintain their vigor and competitiveness. Here are some tips on burning warm-season grasses:

- The best time to burn to increase vigor and production is when they've just begun to grow and have one to two inches of new growth, usually early to mid spring.

- A late spring burn of prairies or native warm-season grasses reduces competition from cool-season grasses such as Kentucky bluegrass, annual bromegrasses (cheat), and smooth bromegrass.
- It's most productive to do burns during two to three consecutive years during a five- or six-year period.
- Prescribed burns at least every three years help control woody vegetation.
- If you'll be applying fertilizers to grasslands, burn first to control weeds that might respond to the fertilizers.
- Burning can reduce nest predation and increase deer browse.
- One- and two-year-old burns produce more food in the way of seeds and also expose more insects for quail than older burns.
- It's a good idea to burn only a fourth to a third of an area in any one year to ensure adequate nesting cover for ground nesters such as quail and prairie chickens.

Burns to Eliminate Brush and Deciduous Trees

Buckbrush (coralberry), western ironweed, saplings of honey locust and Osage orange, and red cedar can be controlled by burning. They should all be burned after they leaf out, as burning will kill the tops. It takes two to three burns to kill all of these except the red cedars; for red cedars one burn usually suffices.

Burning Hazards

Carelessness with fire can damage or destroy property and injure or kill livestock and wildlife as well as humans. You are legally responsible for damages caused by a prescribed burn, including smoke and fire damage. If you intend to burn in an area that has the potential for creating fire damage, first see your lawyer.

Although the damage caused by a fire itself is obvious, less obvious is the damage caused by smoke. Smoke can create serious health and visibility problems. Several conditions must be met to help alleviate smoke problems:

- There must be a wind of at least five to fifteen miles per hour.
- There must be cloud cover less than seventy percent along with a ceiling of at least two thousand feet to allow the smoke to rise above the ground and then be dispersed into the atmosphere.

As the fire produces heat, the updraft helps lift the heat. Hot, fast burns produce less smoke than cool, slow ones. Normally the tall, thick warm-season grasses produce less smoke than cool-season grasses. Fescue produces the most smoke.

Dos and Don'ts of Burning

- **Don't burn within one mile of an airport.** In fact, this may be illegal in some areas.
- **Be aware that smoke can create serious visibility problems for motorists on public roads.** Always burn when the wind is blowing away from public roads, and have enough people on hand to help direct traffic if the wind shifts. You are responsible for a accidents caused by your prescribed burn.
- **Avoid having smoke drift** onto residences, businesses, or farm operations such as dairies, hog operations, horse stables and lots, and chicken houses.
- **Don't stand under or near power lines,** and avoid having heavy smoke go under them. The reason for this is that heavy concentrations of smoke can conduct electricity. Smoke consists mainly of carbon particles and water vapor, which can allow a discharge from a power line to the ground, much like a lightning discharge. Firefighters have been killed in such instances.
- **Don't burn near power-line poles.** They're usually chemically treated and can burn intensively and become extremely difficult to extinguish once they ignite. Spraying water on the poles can create a serious electrical hazard. Fire not only burns poles, but it can also attract an electrical charge.
- **Don't conduct burns at night.** As the air cools during the night, it forces smoke lower to the ground, which can degrade visibility.
- **Before the burn, inform the local fire departments, sheriffs' offices, forestry offices, other necessary public officials, and your neighbors of the location, date, and time of the burn.** This can prevent unnecessary runs by fire departments, and also puts them on the alert should you have to call them in. When the burn is finished, notify all of these people that burning has ceased.

Worker Safety

- **Don't allow people with health problems, including respiratory diseases, heart conditions, and high blood pressure, to participate in burns.** Watch fire personnel carefully for heat stroke, heat exhaustion, and smoke inhalation.
- **Require burn personnel to wear proper safety equipment.** This includes clothing of natural fibers such as wool or cotton that covers the arms and legs (long pants, shirts, and hat or cap)—there should be no

shorts and no short-sleeved shirts. Synthetic fibers melt at high temperatures and can cause serious burns. Other required equipment includes leather gloves and leather high-topped boots. If you're burning near trees or brush, workers should wear hard hats and goggles. Workers operating leaf blowers must wear hearing protection—a logging helmet with hearing protection and a hard-hat is a good choice.

- **Each person working the burn should know basic firefighting techniques.**
- **Everyone should know the burn plan, and escape routes should be planned for every individual.** Everyone should also know the proper procedures for notifying emergency personnel.

Burning Equipment

Prescribed burns require the proper tools. Make sure all equipment is in good working order before beginning the burn.

Sprayer capable of 125 psi pressure and an output of at least six gallons of water per minute (gpm). A tank sprayer on an ATV or on a trailer pulled by an ATV can be extremely valuable.

It's important to have the proper equipment on hand, and know prescribed burning rules. Attending a "burn" school is a very good idea. PHOTO COURTESY ATV COUNTRY, INC.

Backpack blowers. These are useful for firebreak preparation and for putting out fires. The blowers are used to blow leaves and debris from the firelane or fireline.

Hand tools such as rakes, broom rakes, fire swatters, and even wet sacks. All of these can be used to control fire lines.

Drip torch. A drip torch is a tool that holds a flammable liquid such as diesel fuel. The fuel is lit and, as you walk, the ignited liquid is allowed to drip out to start a line of fire. This tool makes it quicker and easier to string backfire lines.

Pliers and bolt cutters. Use them for cutting fences or unlocking gates if you need to escape.

Water. Have plenty on hand with sprayer tanks on trucks, trailers, or ATVs.

CREATING A BURN PLAN

The most important element in prescribed burning is creating a burn plan. The plan should include what will be burned, how and when the burn will take place, and a list of the necessary precautions to take.

Use an aerial photo, topographical map, or even a hand drawing of the area to be burned. Draw or mark in all important features including power lines, gates, fences, houses, and neighboring property lines. Mark the direction of preferred smoke dispersal.

Mark natural or created firebreaks.

- *Natural firebreaks.* Mark streams, crop fields, little-used farm lanes, and closely grazed cool-season grasses. Firebreaks must be wide enough to prevent sparks or burning embers from drifting to unburned areas and starting another fire.

A burn plan must be developed. Firebreaks must be established before the burn. The short grasses and disked area alongside this warm-season grass field is an example of a firebreak.

- *Created firebreaks:*
  - *Bare soil or cool-season grass firebreak.* Create firebreaks of bare soil or cool-season grasses well in advance of the burn. They are usually constructed at least twice as wide as the height of the vegetation to be burned. Bare soil firebreaks are the best, but they can be costly to create and can cause erosion problems. Cool-season grass firebreaks are called "greenlines." Maintain them by by grazing or mowing to keep a short, thick stand of cool-season grasses that has little dry fuel to burn.
  - *Burned firebreak.* Burned firebreaks are established around the perimeter of the area to be burned in areas that aren't bordered by a natural firebreak. Create them by lighting short lengths of vegetation on the downwind side of the burn-area boundary and allowing them to burn about fifteen to twenty feet before they're extinguished. Continue burning short lengths along the perimeter until firebreaks surround the area to be burned.
  - *Wetline.* This is the most common type of firebreak used in prescribed burns. Create a wetline by using a high-pressure sprayer to dampen a narrow strip of vegetation from which a backfire is lit. The fire is lit at the dry area next to the wet strip. Once the fire backs away from the wet strip a foot or so, spray to extinguish it.
- *Heat-reduction strip.* You can create a heat reduction strip by mowing grass adjacent to firebreaks to a height of ten to fifteen inches for eight to sixteen feet. Do not allow mowed materials to windrow. This is a good idea in all cases because there is little fuel to burn and it is easy to control the fire.

Conditions Necessary for Burning

Specific conditions must be met for a safe controlled burn, and weather is the most important. Wind direction and wind speed must be correct to match the burn plan. Relative humidity should be between thirty and sixty percent, temperature should be forty-five to seventy-five degrees F., cloud cover should be clear to seventy percent clear and the ceiling should be two thousand feet or higher. Obtain weather forecasts from local or national sources. Don't hesitate

A prescribed fire is basically a "controlled" fire, using a variety of means including backfiring the downwind side, controlling it, and putting it out on the firebreak. Then fire the upwind side to meet previously burned backfire area. COURTESY ATV COUNTRY, INC.

to cancel the burn if all the conditions are not met. As relative humidity decreases and the temperature rises, burning becomes more difficult to control. Do not burn if the relative humidity is less than twenty-five percent. In many cases you'll have better humidity and temperature conditions in the morning than in the afternoon.

Conducting the Burn

After checking all equipment, discussing the burn plan with helpers and notifying authorities, you're ready to start.

Walk around the perimeter of the area to make sure the wind hasn't blown any fuel across a firebreak.

Begin the burn in a downwind corner of the area. Start with a small test fire to make sure everything and everyone is working properly. Then continue to light the backfire (the fire that is lit on the downwind side, but controlled by a wetline or heat reduction strip) in sections small enough for the burn crew to handle. Continue this around the perimeter until you reach the upwind side. Wind directions dictate the backfires and types of burn. They may be ring burns, strips, head fires, or flank fires.

Assign three or four people to each fireline. One lights the fireline, one or two control the fireline, and one is left to mop up or make sure no fires restart outside the firebreak. An L-shaped fireline requires at least six workers.

Watch the weather and wind and adjust the burn plan as necessary.

Once the burn is completed, recheck the perimeter to make sure sparks or embers haven't created a fire outside the firebreak or area. Snags, brushpiles, cow chips, or other long-lasting fuels near a perimeter can restart a fire. Move them into the burn areas away from the perimeter. Continue to monitor the area. Once you are satisfied that the burn has ceased, notify your neighbors and officials.

Even a "controlled" burn can be dangerous and a liability. We can't describe everything you need to know in one book chapter—entire books have been written on the subject, and burn schools are available. Make sure you obtain the information and schooling you need before you conduct a burn. Consider becoming "burn buddies" with a neighbor or friend who will help you conduct your burn if you reciprocate—and with whom you can share experiences. Many SWCD and state forest personnel will help with or advise on prescribed burns.

CHAPTER 26

WHITE-TAILED DEER

Do you want more deer to hunt? Bigger deer on your land? Bigger-racked bucks? All of these are possible with proper management. First, however, you have to decide which of these is your goal, because management for one is not necessarily the same as management for others.

GENERAL MANAGEMENT FOR DEER

It's fairly easy to create the ultimate whitetail habitat, even on small properties of forty acres or so. Just provide the basic requirements, which include food, cover, and water. Let's examine each of these requirements.

PROVIDING FOOD

Food can include mast (acorns), browse, agricultural products, plants in food plots, supplemental foods, and minerals. Deer are typically browsers; they

Managing for white-tailed deer is one of the most popular forms of wildlife habitat management. Providing a diversity of food, with food plots, timberland management, shrubs, and forbs is the first step.

tend to nibble here, nibble there, and then move on to another plant, so the more variety you can provide, the better. Your management plan should include making sure that food is available all year. Providing food can be done in several ways, depending on the native fauna and existing agricultural foods available, how much time and money you want to spend, and local laws.

Managing woodlands for food production. Managing woodlands to increase mast production and/or growth of native forbs is a great way to attract deer. This may involve selective timber cutting, the use of prescribed burns, and the use of selective herbicides, all of which can open up a mature timberland to produce more forage. Ask your state or local forester for a plan.

You can also fertilize selected mast trees to get better production. A good mixed agricultural fertilizer such as 10-10-10 or 12-12-12 will work, although commercial tree fertilizers are also available. Apply approximately a pound of fertilizer per inch of tree diameter (at breast height), spreading the fertilizer around the tree in a circular pattern to the edge of the drip line.

Increasing nutrient concentration. There are several ways to provide deer with increased nutrients, and they can be done separately or together.

- **Create food plots.** This is the most widespread method of increasing nutrients available to deer. Food plots planted in perennials, including legumes such as the clovers, provide many nutrients in a small area. Food plots may also consist of annual winter foods such as wheat or winter rye or annual agricultural crops such as corn, soybeans, or milo. You'll find many seeds suitable for food plots locally or from specialty seed companies, some blended or marketed just for whitetails. All of my food plots contain a variety of perennials and annuals for year-round use.
- **Provide supplemental feed.** This is considered baiting. Some states don't allow it, some regulate it strictly, but it is allowed in still others. Supplemental feeding should not be used simply to attract deer, but should be a part of a total management plan to create the ultimate habitat. Supplemental feeding should also not be emergency feeding (feeding after a really deep snow or during a drought). As it takes deer some time to shift their diet intakes; emergency feeding can actually be harmful to them because they can't shift their diet that quick, and if you stop, they may "starve" even

with other food available. You can provide grains, such as corn or soybeans, or commercially produced pellet feeds as supplemental feed. Corn by itself, however, is not a good winter feed as it's relatively low in protein—only seven to nine percent. The amount of protein in a deer's diet can make a big difference in its health. Missouri researchers reported a forty-two percent loss in fawns from does on a seven percent protein diet, a twenty-seven percent loss on a ten percent protein diet, and no losses on a twenty-three percent protein diet. Supplemental feeds can be hand-distributed, distributed in electronically controlled feeders or gravity drop-style feeders, or even in feed bunks.

- **Provide minerals.** Minerals are also considered bait in some states, and some products definitely are baits, while others use scent and taste to entice whitetails to consume the minerals and vitamins. A Pennsylvania study concluded that dietary levels of 0.64 percent calcium and 0.56 percent phosphorus were needed for optimum antler growth. Unfortunately, even the best natural habitats cannot supply these minerals naturally; they must be provided as supplements. Many minerals formulated just for whitetails have become available in the past few years. Before their introduction I used a cattle salt/mineral mix to which I added dairy calf supplement. (See Minerals and Supplemental Feeding.)
- **Grow native shrubs.** Plant, fertilize, or encourage shrubs, forbs, and vines so they produce more food. Soft mast, such as dogwoods, grapes, apples, and persimmons, is often overlooked and can also be introduced or managed. Apple trees are a favorite of deer, and they're now available in wildlife bundles for planting from state and county agricultural agencies.

Providing Cover

Deer are basically "edge" creatures. They don't like mature timber, nor do they really like wide open country without any cover. The more edge you can create, especially around food plots, feeders, or travel lanes to these areas, the more likely it is that the animals will stay on your property or near the food. This means that you should provide bedding areas nearby as well.

Providing cover near these food sources is also an important factor in whitetail deer habitat practices.

If you have mature timber, the selective cutting you do to encourage growth of more food can also be used to create small clear-cuts that quickly resprout into thickets suitable for bedding. You can increase the vegetation in these thickets by fertilizing and encouraging or even planting vines such as wild grapes or honeysuckle. If you plant honeysuckle, though, plant only the native vining species, not the exotic bush honeysuckle. The latter is an exotic invasive plant that has no value and spreads rapidly.

Food plots alongside wooded areas do best if there's edge between the two. Plant shrubs, soft mast, and vines along these travel corridors.

Deer can withstand winter quite easily without a lot of heavy cover in many parts of the country, but they'll select thick cover when the going gets tough. Leave or create some thick evergreen (cedar or pine) bedding areas. You can even partially cut some evergreens, allowing them to fall over but still remain alive for deer bedding cover.

If your property doesn't have timber, you may wish to create woodlots. It will take a few years, but fast-growing oaks such as the sawtooth variety can provide food and wooded cover with a little patience (about ten to fifteen years). Travel lanes and small woodlots provide better habitat for deer than large blocks of timber.

If your property is mostly open, you can encourage the growth of cover by introducing native warm-season grasses. Deer love to bed and hide in these grasses, as anyone who's hunted in Kansas can attest. Encourage brushy cover draws, and manage and enhance riparian corridors.

Water is also an essential. Small wildlife watering ponds located in timberlands or near food plots will attract deer in areas without other water sources.

Providing Water

Water is a necessity for a deer's survival, although plans to provide it are often overlooked. Deer will use any water source they can find. In fact, hunting around stock tanks is a common tactic in Texas, where water is often scarce. Farm ponds, lakes, rivers and small streams, and even spring seeps are all sources of water.

Water is the final key to creating the ultimate whitetail habitat. I've created two wildlife watering ponds in areas of my property where water was not previously available, and the daily tracks around the ponds tell the story.

Evaluating Your Location

It's a good idea to consider your property as a total whitetail habitat. Place the major food and bedding areas in the center of the property, away from neighbors, public roads, and areas where other hunters may have access to the deer herd. This will help keep the deer from wandering onto neighboring property, although of course you can't prevent a buck in rut from traveling anywhere. I've found it best to place all the food sources in one "cafeteria" area. I've also placed a food area in each forty-acre parcel of timber on my property. One reason for this is that after years of research on does, I have found that there are established home ranges for different doe groups at each of these locations. The numbers and locations of the food areas will vary according to your property size, shape, and location.

It's important to establish complete habitats that doe families can live in year-round. Although the bucks don't associate with the doe groups, believe me, the bucks will find them. With all the food and cover, you'll have bucks living on your property year-round as well.

If you provide these essentials and manage them properly, you will have deer—lots of deer—and possibly a few big bucks. Attracting the really big ones, however, takes more management. You'll have to consider two more factors if you want big racks—letting the small bucks walk (i.e. don't shoot them), and managing your property to keep doe population levels in check. On our property that means that for every buck taken, the hunter must take a doe.

Managing for Doe Groups

Hunting does is a tactic for taking bucks that's been used successfully for a long time. You can also manage your property for doe groups in such a way that you have more opportunities to take bucks visiting the does. Doe groups have specific requirements, and in many cases it's fairly easy to manage an area

Although most hunters "manage" for big bucks, the best tactic is to manage for doe groups or doe families. This attracts the bucks and keeps the big, dominant bucks on your property.

to increase the number of different doe groups rather than simply managing for "more" deer.

Although we see does and bucks together quite frequently during the rut, for the majority of the year bucks and does live in separate groups, and often they live in separate areas. In early fall, however, yearling groups may be made up of both sexes. Throughout much of the year younger bucks stay in loose bachelor groups, while the older, dominant bucks stay pretty much to themselves. The doe groups stay pretty much to themselves as well. They are matriarchal, made up of mothers, daughters, grandmothers, cousins, aunts, and so forth. One of the reasons for this segregation is that each group has different needs. Does tend to have small home ranges, and the size of the range depends on the quality and type of habitat. Bucks, on the other hand, have larger home ranges and tend to roam more during the entire year.

Typical doe-group home ranges, on good habitat, can range from forty acres to a half-mile square (160 acres). The size of the home range is larger in areas with less than ideal habitat, or where food and cover for bedding areas are widely separated. In addition, each doe of breeding age will have her own "safe" fawning territory, which biologists have determined is about twenty acres on ideal habitat. She will chase other does, including her own young of the year, away from this area during fawning time. Other than during the fawning time, there's usually a lot of overlapping of doe territories, as there is with bucks. Bred yearling does will often set up their territory next to their mother's. They will, however, also have their own distinct fawning area.

These doe home ranges, and in particular the fawning areas, have specific requirements, and in order to increase the number of doe groups (not particularly the numbers of does), property must be managed to provide the availability of separate "zones" for several doe groups. Theoretically, you could manage a two hundred-acre property for four doe groups if the right food and habitat requirements were met and other management practices were followed, including harvesting does to keep the buck/doe ratio more even. This is not always possible due to geography, soil types, land uses, and so forth. I have, however, accomplished that goal on my property with intensive management. The requirements for doe groups are food, water, and cover, all within a small core area. Fawning areas especially need areas of heavy cover.

Assessing Your Property

Use an aerial photo to determine existing habitat, boundaries, places where it is feasible to do management, and places where it is not. For instance, creating a food plot along a highway, on the edge of your property near adjoining property, or near houses is not a good choice. Determine water sources. If water is not available, creating small watering ponds may increase the attractiveness of some areas. Of course food is important, and it can be a mixture of native and planted foods. Although all of these ingredients are important, the most important is cover, particularly in the fawning area. It should be brushy—almost predator-proof—yet with open areas nearby. The brushy areas don't have to be big. I've created several thickets as small as an acre on my place, and invariably when I go to pick raspberries and blackberries from these areas during the early summer months, I jump or spot fawns. Determine the needs of several possible doe zones on your property, and then manage to create or increase those portions of the habitat that are missing from them.

Creating bedding areas and refuges that are not only not hunted, but left entirely alone with no visitation by humans, can also help keep big bucks on your property.

Creating Cover

On many pieces of property managed for deer, the prime habitat is woodland. Open woodland provides mast, but little cover. Small brushy tangles interspersed throughout the woodland offer the cover deer need. They can be created by simply clear-cutting a small area (less than an acrea) and not treating the stumps with herbicide but allowing them to sprout. This provides not only cover, but also browse, a favorite deer food. If the sprouts become too big, cut them every few years with a brush saw and they'll regenerate. In most cases these areas will quickly grow into a tangle of sprouts, weeds, forbs, and vines. It may take a year or two, but you can shorten the time by planting blackberry briars, raspberry vines, honeysuckle, soft mast trees, shrubs, other vines, and even native warm-season grasses. The grasses offer high vegetation to conceal fawns from predators, and growing them has produced a marked increase in fawn survival rates in studies done on several ranches in Texas.

Once the plant succession reaches its later stages and trees again begin to dominate, simply remove them. Cut them off approximately six inches above ground level and again allow them to sprout. Adding a fertilizer such as Scotts Native Plant Fertilizer can increase the productivity and shorten your thicket's growing time.

Creating or Providing Food

Creating food is relatively easy, but it's important to provide food sources year-round, and particularly important to provide a good food source in the winter and early spring months. This not only provides the nutrition when the deer herd needs it the most, but it also tends to hold the does in their ranges just before fawning time. You can provide natural foods—hard mast, soft mast, browse, and forbs—as well as food plots and, in areas where it's legal, supplementary feed.

Natural foods. Natural foods often occur in timberland, and most states have forestry experts in their fish and game departments who can evaluate your property and recommend timber management practices that can increase food and cover for deer and other wildlife. Often simple TSI practices, such as thinning and culling of non-productive trees, are all that's needed. Following these practices removes competition from the productive mast-producing trees and encourages undergrowth of browse. Specific areas of timber can also be treated with BASF Habitat Release to kill brushy plants and encourage the growth of forbs. Prescribed burns can also be extremely helpful in creating food sources in timbered areas, especially in areas where the predominant habitat is even-aged brush. Soft mast, apple trees, vines, and shrubs preferred by deer can also be planted in out-of-the-way places to provide more food—around the edges of food plots, along farm or logging roads, and around the edges of overgrown fields.

Food plots. Once natural food sources have been created or increased, consider food plots. They can be used to create year-round food sources if you choose the right plants. If clearings for food plots do not exist, you can create them in timbered areas by dozing, chainsawing, or brush-clearing, and then planting. Food plots can produce almost immediate results and can consist of annual or perennial plants or a combination of both. Clover and alfalfa are popular choices because they are fairly easy to plant and are perennial. They provide excellent nutritional sources and are a favorite deer food. Food plots should be at least an acre in size, although several smaller plots interspersed in smaller areas may be even better as they provide more edge. However, they are harder to pattern and hunt. Annual food plots can provide excellent winter food sources. Corn, milo, and soybeans are excellent choices, and oats and wheat can also be used, with oats the best choice. Have a soil test taken before you plant and adjust the pH of the soil with lime and bring the soil up to test with fertilizer before planting.

Supplemental feeding. If it's legal in your area, supplemental feeding can be used to provide more nutrition during the winter months. Dan Moultrie of Moultrie Feeders says his favorite supplemental feed is soybeans because they contain more protein than corn. He starts out with corn, which appeals to deer, and gradually switches over to soybeans. A number of commercial "deer pellets" are also available, although they take more effort in feeding. (See Chapter 18, Supplemental Feeding.)

Salt/mineral supplements. One of the most important steps to take in producing a healthy deer herd, and the final step in creating a doe zone, is providing salt/mineral supplements. At least one mineral lick should be provided for each doe zone. In my camera/monitor studies I've photographed numerous does on each of a half dozen mineral licks at approximately the same time. "Deer minerals" are available commercially, and the better ones also have trace minerals and vitamins. These supplements should be provided free-choice on a year-round basis to provide for the pregnant and lactating does as well as for antler growth of the bucks. (See Chapter 17, Essential Vitamins and Minerals for Wildlife.)

It's important to keep food and supplements grouped together in the separate doe-use zones of your property. With proper management for does in each areas you can spread out the deer herd more, provide more food for more deer and increase the chances for taking more bucks.

Provide Bedding Areas

Two days after the deer-hunting season ended one year, curiosity got the better of me and I sneaked into a bedding area. I had taken a decent buck between the bedding area and a food plot, but I knew there had to be a better buck using the area. I had seen a huge one a couple of times while bowhunting, but never got a shot. Sure enough, there was a giant fresh footprint—I'd have a big buck to hunt the following year. Food plots can be a great help in attracting deer to your hunting area, but they are not the only key. Having a good bedding area nearby is just as important. If you have the food but the bedding area is on your neighbor's property or some distance away, you may not get a shot at the trophy bucks using your food plot because they may not come onto your property until after dark.

If you don't already know the location of bedding areas on your hunting property, scout for them. Deer choose secluded, heavy-cover areas for bedding. The most obvious indicator of a bedding area is a large number of droppings scattered around but still fairly close together. The first thing a deer does after getting up from a bed (assuming he's undisturbed—not startled) is to stretch, move about a bit, then urinate and defecate. You can also sometimes spot slight depressions in vegetation or leaves created by bedded deer. Unless they are disturbed or food patterns change, deer tend to bed in the same general areas. Late winter, after the hunting seasons are over, is a great time to scout for bedding areas or for places to create bedding areas. If a covering of snow is on the ground, scouting is even better. Use a topographical map and an aerial photo of your property to help locate potential bedding areas. Mark them on the map and then scout to confirm.

If you don't have bedding areas, create them. I made my first bedding area over twenty years ago by accident. My farm was a former Ozark hillside dairy, and the timber had been allowed to grow into a mature forest. The farm had also been grazed down to practically bare rock, even in the wooded areas. You could literally see bare ground for several hundred yards through the timber. There were no deer on the property at that time. I selectively logged the farm, and in two spots I clear-cut areas of about an acre. At the time I wasn't trying to attract deer—the trees were all old and dying and it just made sense to take them all out. Within two years the clear-cuts had sprouted into a maze of shrubs and saplings, including blackberry brambles, buckbrush, grape vines, and young oaks and hickories so thick you almost couldn't push your way through. Many of the selectively logged areas had improved as well. These areas attracted deer like magnets. One of the clear-cuts was near an old overgrown field I had rejuvenated as a food plot with ladino clover. The first year a stand placed between the bedding area and the food plot produced our first deer, and it has produced deer every year since. Creating your own bedding areas can be done in several ways depending on the existing vegetation on your property.

Choosing a location

You may already have a bedding area or a place that needs only a little more management to make a good bedding area. Ideally the bedding area should be fairly close to but not right on the food plot—a hundred yards or so from a food plot works well. The best locations are secluded, although deer will bed quite close to human activity if it's the only available area. The location should also be away from property boundary lines and public roads. The beds deer use most often are located near the tops of hillsides or draws, with an escape route over the top of the hill. Deer, and in particular big bucks, like bedding spots that face into the prevailing wind but that also have a quick escape route downwind. Other excellent bedding areas are creek and river bottoms, as these typically provide everything deer need to survive—water, cover and food.

Types of Bedding Areas

The types of bedding areas depend on the terrain and prominent vegetation in the area. High grass and even uncut corn make great bedding areas. Even shorter vegetation, such as clover fields, offers nighttime bedding areas for deer as they feed, lie down, and then get up and feed a bit more.

For the most part, however, bedding areas consist of extremely thick cover that offers lots of protection with quick escape routes. Brushy draws offer daytime hidey-hole bedding areas in prairie or heavily farmed country. And the area doesn't have to be big. A hunt with Harold Knight and David Hale several years ago revealed this fact. Harold had seen a buck stand up in a tiny brushy draw, no bigger than twenty yards at its widest, a couple of times during his day-before scouting. The draw was surrounded on all sides by a bean field with no place for the buck to go—or so it seemed. I stood at the bottom of the draw as David and Harold did a drive from the top toward me. When they were about halfway to me, I noticed movement behind them and with my binoculars spotted a buck with a massive rack belly-crawling to the top of the hill. He had simply let them walk by and was now sneaking away.

Here are some ideas for making bedding areas in draws:

Simply let existing brush grow in draws. Brushy bedding areas are easy to maintain and create in draws as well as odd corners of fields, and hard-to-get-to areas.

Plant native warm-season grasses such as big bluestem, little bluestem, switchgrass, and Indian grass in a fifty-foot wide strip on either side of a draw. The native grasses grow tall, but in clumps, leaving nice open bedding areas, and plenty of height to hide even the biggest buck. The native warm-season grasses also offer some foraging alongside the bedding areas, as well as providing erosion control for the edges of the draws.

Plant native shrubs if a draw has little in the way of cover; wild plums are an excellent choice in many prairie and farmland regions.

Most bedding areas in the South, Midwest and Northeast, however, are thickets of some sort, either in wooded areas or near them. My good friend the late Ben Lee, while driving on a major highway to a hunting spot, would point to an Alabama roadside thicket and comment, "You couldn't guess how many deer are bedded right there right now." If you create a thicket, you'll create a bedding area, but creating a thicket takes a little planning and depends on what type of wooded vegetation you have. Here are some tips for creating thickets.

Contact your local conservation agent or state forestry department for advice and low-cost plant bundles. Many states also have wildlife improvement programs through local soil conservation offices located in the county seats. Both types of agencies have personnel and sometimes even money or plants for wildlife management projects. Missouri offers lots of advice and low-cost wildlife plant bundles consisting of shrubs, trees, and other plants that can be used to create cover for wildlife. Missouri also offers the Wildlife Habitat Improvement Program (WHIP) for landowners, and my farm is enrolled in that program. Often these cover-for-wildlife programs and their resulting "projects" are the makings of great deer bedding areas.

Grow native plants from seed collected from plants in your area. This takes a bit of expertise. Ben Lee had a great habit of carrying a pocketful of honeysuckle seeds and planting them when he was scouting or spring turkey hunting. Honeysuckle quickly turns almost any area into a deer haven. The honeysuckle must, however, be the native vining plant instead of the shrub honeysuckle sold through most nurseries. The latter is an exotic that can spread and compete with native plants.

Consider cedar thickets. In the Ozarks where I live, cedars are often allowed to grow in selected areas to create thickets or bedding areas. Cedars grow into thickets a person can't crawl through but bucks love. They do, however, quickly spread to fields and other areas, and it's important not to create problems in your attempts to create bedding areas.

Selectively thin a timbered area. This may be the simplest and best way to create bedding areas, especially in some old-growth forests. Again, advice from your state forester can be invaluable. TSI often involves selective thinning of less valuable trees, leaving the more valuable trees, including selected mast trees. It's important to choose the correct trees for thinning, leaving the best mast-producing trees, but in an uneven age grouping. This opening of the top canopy creates an instant explosion of understory and results in a bedding area without having to clear-cut. In fact, an acre or two of TSI timber can provide lots of bedding areas. These timber-managed areas provide not only bedding areas for deer, but food for wild turkeys and grouse as well. The trees from the thinning can be used for firewood, or possibly sold as logs. An alternative to cutting down the trees is to girdle selected trees by cutting through the outer layers completely around the tree and applying a herbicide such as BASF Habitat Release to the cut.

Clear-cut small areas to create bedding areas. This is usually not necessary and often results in an overabundance of dense sprout growth. If you have a fairly large amount of wooded acreage that needs TSI for wildlife, you can work on a few acres each year. Doing this provides a continually changing and usually improving timber for wildlife such as deer.

For the most part bedding areas don't have to be large. Even a quarter acre of the thick stuff will harbor deer. A bedding area located in the middle of pure open woods, with no escape routes, is not as effective as one with cover routes created from the bedding area to feeding areas. As with food plots, it's best to have a number of smaller bedding areas scattered around your property than to have one large area. Don't allow livestock to graze these bedding and TSI areas—if you do, you'll defeat your purpose.

With proper habitat management deer will flourish. The last step in quality deer management is to let the little bucks go and aggressively hunt the does to maintain a balance and prevent overcrowding of the area.

Don't Hunt Bedding Areas

The final key to success with bedding areas is to avoid hunting in them. For the most part does and young deer will still use the areas even if you hunt them, but big bucks will quickly depart for safer country. Use the bedding areas as a refuge. If you provide food plots, bedding areas, and good travel lanes between them, you'll have created your own deer haven, even on a small property.

CHAPTER 27

WILD TURKEYS

Wild turkeys are America's wildlife success story. This noble bird was once found in large numbers in the United States. "Too many wild turkeys even to consider raising tame birds," was a quotation from the late 1800s. By around 1925, however, the wild turkey was becoming scarce. Starting with a remnant population in the mid-fifties, wild turkeys were reintroduced by live trapping and relocation to create the burgeoning population that now covers much of the United States. The wild turkey has become the second most sought-after game animal after the white-tailed deer.

HABITAT MANAGEMENT

To manage habitat for wild turkeys, you must develop the combination of food, cover, and water that will produce the maximum number of turkeys compatible with other land uses. Vast acreages of timberland were once supposed to be the ideal wild turkey habitat, but the fast growth of turkey flocks in northern Missouri, with a habitat that's mostly woodlots and vast grainfields, has changed people's minds about that. The reintroduction of Merriam and Rio Grande turkeys in the Plains states has also emphasized changing management practices.

Next to white-tailed deer, the next most popular American game is the wild turkey. Habitat management to produce maximum numbers of wild turkeys on the land can involve a number of practices.

And turkeys have learned to adapt quite readily to proximity to humans. We see turkeys around our home almost daily, and they strut almost within the city limits of a small town near us as well. Turkeys have also adapted to cattle. When our cattle are rotated out of a pasture, the turkeys quickly move in and flip over the cow chips to get at the bugs under them.

Many of the practices recommended for white-tailed deer in the previous chapter will benefit turkeys and, in fact, if you have a growing deer herd, you probably also have a growing turkey flock.

There are, however, some differences between what turkeys need and what deer need. Wild turkeys need three distinct habitat types annually, depending on the sex and age of the birds—winter, nesting, and summer/fall.

Winter Habitat

The most important part of the bird's annual range is their winter habitat. Wild turkeys normally spend about six months (October through March) in winter habitat, which must provide a reliable and adequate food supply and cover during bad weather. Acorns are the staple of a turkey's winter diet, so high-quality winter habitat should be at least fifty percent timbered with oaks—several varieties of oaks—that are large enough to produce acorns. A TSI program often helps to provide more food. In agricultural country, grain crop residue substitutes for mast in a turkey's winter habitat.

Winter habitat, consisting of woodlands with a good supply of hard mast, is the backbone of a successful wild turkey management program.

Nesting habitat is extremely important. This can be the edges of old fields, loose brush piles in clearings, or along trails. Turkeys usually nest fairly close to a permanent water source.

Nesting Habitat

The turkey's nesting habitat can vary greatly. Hens usually nest around the edges of old fields, in hay fields, in patches of blackberry briars or similar cover, in loose brushpiles, along woods trails, and at the edges of wooded clearings. Most turkey nests are fairly close to a permanent source of water.

Once a hen turkey has mated with a tom, she begins laying eggs at a rate of one per day until the nest contains ten to twelve eggs. At that point, she begins incubating the eggs, which take twenty-six to twenty-eight days to hatch. Hens generally nest where ground cover is dense yet open enough for the hen and poults to walk through and short enough for the hen to see over.

On average, probably only twenty to forty percent of all turkey nests are successful. In bottomland areas, many nests are lost each year due to flooding. In addition to the destruction of nests from flooding and other extreme weather conditions, nest predation (by coyotes, bobcats, raccoons, opossums, skunks, feral dogs or cats, black rat snakes, and humans) also takes its toll.

Small predators such as raccoons, opossums, skunks, and black rat snakes destroy nests by consuming and/or breaking the eggs. Coyote, bobcat, and human predation often results in the hen being killed as well. Humans are included in this category because countless hen turkeys are killed or injured every year while sitting on their nests because they're run over by a tractor and bush-hog or other mowing equipment.

The loss of turkey nests as a result of weather conditions and predation by animals other than humans is probably uncontrollable for the most part. There are, however, a few preventive measures you can take to significantly reduce the number of nests destroyed by humans.

Refrain from mowing or bush-hogging fallow pastures, overgrown fields, fire lanes, or roadsides until the end of June. (If a nest is destroyed during the first egg-laying period, the hen may attempt to renest up to three times. This translates into turkeys sometimes sitting on nests up to the end of June.)

After the eggs hatch, even though the poults can walk, they can't fly until they are about ten days old. Until then, they are susceptible to being overrun by machinery.

If you accidentally flush a hen turkey from the nest, leave the area immediately without disturbing the eggs. If the hen has begun incubating the eggs, she'll probably return to the nest as soon as you leave.

Summer and Fall Habitat

Summer and fall habitat is used during late summer and early fall, the time of year when hens and poults are seen more frequently in open areas. These areas may include mowed hay fields, grazed pastures, glades, areas alongside logging roads, food plots, and even open woodswhere the young birds have the opportunity to capture insects and browse on leafy vegetation. Food plots created primarily for deer, using plants such as clovers and alfalfa, which attract large numbers of bugs, are great summer bugging areas for poults.

In comparison to that of winter habitat, the size of the summer and fall range used by turkeys is relatively small. This range is, however, extremely vital. The acreage in openings may vary, but it should be more than ten percent of the total annual range; thirty to forty percent is the optimum. As with deer, edge is also extremely important. It's good to have a variety of plants there.

To ensure a dependable source of natural food for turkeys (and future timber supplies) strive for an equal distribution of age and size classes of trees on your timbered lands. Approximately one-third should be in pole-sized trees (two- to nine-inch DBH) and one-third in mature saw logs. When this sort of balance is achieved, a dependable mast crop will be ensured, and the openings created when stands of saw logs are harvested is an added benefit.

Turkey Foods

Turkeys will remain at a food source, no matter where they have to go, and they'll eat a wide variety of foods. With abundant food sources in spring and

Turkeys thrive on a variety of foods including plants and insects. Creating openings in timber adds diversity of foods. A management plan should be able to meet year-round food requirements.

summer, they stay in relatively small areas. When food is scarce in winter, they may range a great deal.

The principal foods turkeys eat fit into a few general categories: hard mast (oaks, hickories, walnuts, pines); soft mast (grapes, cherries, gums, persimmons, junipers, dogwoods); seeds (native grasses and sedges, corn, oats, weeds); and greens (grass and grass-like plants including annual and perennial forbs). Not only do they eat a variety of plant species, they also eat a variety of plant parts—fruits, seedheads and seeds, roots and tubers, stems, buds, leaves, pods, and capsules and flowers.

Wide fluctuations in the supply of one type of food, such as hard or soft mast or seeds, usually are not critical to turkeys because low production in one category usually coincides with high production in another. The different oak species produce their maximum seed crops during different periods, so in effect their production periods alternate and ensure a yearly hard mast crop. For instance, white oaks produce mast annually, but black oaks produce biannually. Many fruiting soft-mast crops, such as wild grapes and dogwoods, produce more heavily every other season.

Agricultural crops such as corn, soybeans, cowpeas, sorghum, millet, oats, wheat, buckwheat, and peas, can also be an important part of the wild turkey's diet, and all of them are desirable. High populations of deer, however, can compete with turkeys for them.

Design a habitat management plan that provides year-round food for turkeys using the following lists, which show preferred foods and the times they're available.

FALL AND WINTER FOODS

Crabgrass seeds (fall)
Grasshoppers (fall)
Wild grapes
Dogwood fruits
Tick trefoil
Sheep sorrel
Panic grass seeds
Lespedeza seeds
Greenleaf material (clovers)
Corn and other crop residues
Hard mast

SPRING AND SUMMER FOODS

Grass leaves
Greenleaf materials (clovers, alfalfa, brassicas)
Sheep sorrel seeds
Grasshoppers, beetles, and other bugs
Bluegrass seeds
Crabgrass seeds (late summer and fall)
Blackberries
Currants and other berries
Mushrooms
Acorns (early spring)
Panic grass seeds

FOOD MANAGEMENT PRACTICES

Acorns are the number one choice of food throughout the year, but grasses, sedges, and agricultural crops can also provide foods at certain times of the year. Here are some basic food management practices.

TSI. The same tactics used to improve timberlands for whitetails is also effective for providing more hard mast and soft mast for turkeys. Turkeys do, however, prefer more open timberlands. At least one-third of the timberland should be in mature stands with little undergrowth, and at least ten percent must be in clearings or openings.

Food plots. Annual grain food plots for turkeys can supplement natural food supplies. These food plots can be especially helpful in times of extremely bad weather or during drastic shortages of natural food supplies. Food plots should have a mixture of grains, including corn, soybeans, sorghums, and a strip of winter wheat. Soil test, fertilize and lime as needed.

Regardless of all else, the primary food for wild turkeys is mast. This is especially important during the winter.

Chufa. Planting the annual chufa can attract, hold, and feed turkeys. Turkey Gold Chufa, sold by the National Wild Turkey Federation, is an agricultural variety of this native nutsedge that matures in 90 to 110 days. Chufa produces tubers that turkeys relish. They begin scratching the tubers out during the fall months and will continue to feed on them until they're gone. PlotSpike Chufa from Ragan & Massey is also available.

Chufa grows best in sandy, loamy soil. Fertilize and lime according to soil test. For best results, plant chufa in a well-prepared bed after the final frost. Plant seeds no deeper than one-quarter inch and supply adequate moisture to assure germination. Plant ten pounds per quarter acre.

Green browse plots. More-permanent food plots consisting of ladino clover or a combination plot of orchard grass, ladino clover, and Korean or Kobe lespedeza can provide green materials and bugs, and gobblers love to strut in these areas. Soil test, lime and fertilize. Apply no more than twenty pounds of nitrogen per acre to avoid excessive vegetation growth. Turkeys prefer thin stands of vegetation and may not use dense, lush stands. Prepare the ground and seed in the fall with half a bushel of winter wheat and two pounds of orchard grass per acre. Overseed half of the plot in the fall and winter with two pounds of ladino clover and two pounds of red clover per acre. Overseed the second half with ten pounds of Korean or Kobe lespedeza.

Idle fields. Old abandoned fields bordering or surrounded by timber are traditional turkey habitats. They often include former house sites with bluegrass, an important turkey food during spring and summer. Keep these fields open and in a grass-legume mixture. Mowing or moderate grazing improves the quality of these fields since turkeys tend to avoid fields grown up in dense broomsedge.

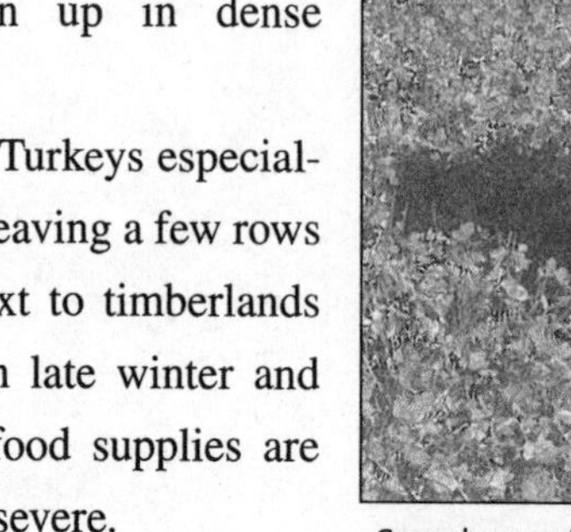

Green browse food plots, primarily of clover, provide green material as well as insects for poults.

Crop residues. Turkeys especially like corn fields. Leaving a few rows of corn standing next to timberlands will provide food in late winter and early spring when food supplies are short or winters are severe.

Supplemental feeding. In the Midwest, where we live, as well as in the West and in northern turkey ranges, wild turkeys gather in big numbers near agricultural feeding operations in the winter, especially when the snow is deep and the weather bitter. These include dairies, cattle feedlots, and pasture feeding areas. We began a wildlife feeding program using shelled corn on our farm many years ago. It has some pros and cons, but after many years of doing it, we feel that the pros far outweigh the cons. Some suggest that feeding concentrates the birds, which can cause greater predation or disease outbreaks. We have not experienced these problems, mostly because we have numerous feeders, so that large flocks are not confined to one area. In my farm travels, I see as many signs of turkey predation away from the feeders as near them. We feed primarily in winter, and we feed in addition to following the other habitat practices I've mentioned. Once spring arrives, we stop feeding and the flocks spread out naturally. There are, however, some dos and don'ts.

- Make sure to keep feed available consistently throughout the winter once you start.
- A good general rate of feed is about a cup of feed per day per turkey. Whole kernel corn, cracked corn, oats, or wheat can be used. Don't dump feed in a pile— broadcast it over a wide area or use feeders that throw it.
- Keep human contact to a minimum. Don't feed close to houses or highways or in areas of high visibility.

WATER

Wild turkeys must have water, and they usually aren't found very far from it. They often like to roost over the water around lakes, streams, and swamps. You can improve turkey habitat by constructing a small water pond where there is no permanent water for each quarter section (160 acres) or for every forty acres.

COVER

Wild turkeys prefer the more open woods. Studies have shown that saw timber stands are home to twice as many turkeys as pole-stage timber stands. Turkeys will use timber stands that have grown beyond the small pole stage if the understory is not too dense. The open understory provides a unique floor litter that produces insects and herbaceous forage desired by turkeys. There may be a psychological need for the open timber as well.

Studies have shown that an extensive red cedar invasion can reduce the turkey carrying capacity of a tract of timberland by as much as fifty percent.

Quality forest management is a must for quality turkey habitat. Forestry practices in the past and those used today influence turkey populations more than any other land use practice. Even-age timber management or clear-cutting has been used for many years in pine forests. It's used less in hardwoods, where selective cutting is the norm. Even-aged management while good for turkeys, however, creates clearings and openings that might not exist with selective cutting.

Turkeys prefer the more open, mature woods. The open understory provides insects and herbaceous foods. Keep at least one-third of the wooded tract in saw timber or larger, mature woods.

CHAPTER 28
UPLAND BIRDS

Upland birds, including quail, pheasants, mourning doves, prairie chickens, and sage grouse, can be both the easiest and the toughest wildlife to manage for. Each has specific needs that must be met.

BOBWHITE QUAIL

Bobwhite quail populations have been in a decline across the nation for many years. When I was a kid growing up in the rural Midwest, quail were the premier sport. Actually, quail hunting was the only hunting my dad enjoyed. Experts point to several factors that probably play a part in the decline of bobwhite quail, including predators, habitat loss, and disease. Although there is no single cause for the decline, habitat loss is the biggest factor because the other two are related to it.

Mr. Bobwhite has been in a decline for many years, primarily due to the continuing loss of proper habitat. Quail have more specific needs than many other wildlife and require more intensive management.

Three types of quail habitat exist—small grain farms, open timberlands with openings, and natural prairies or grasslands. Back in the '40s through the '50s and even into the early '60s, most grain farms in the quail's home range—the South, Midwest and Southwest—were fairly small—at least the fields were small and were often ringed with brushy fencerows, ditches, creeks, or woodlots. Grain harvesting was not very efficient and lots of waste grain was left in the fields. The fields were also often left fallow until the next spring, when they were again worked, and this left even more food. Pasture and haylands were seeded in Kentucky bluegrass, redtop, and lespedeza, all quail-friendly plants that provided seeds and were fairly sparse, allowing the birds to move through the plants yet have protection from predators. Woodland birds had lots of open woods with sparse underbrush that allowed access to seed heads and forbs.

Intensive tree farming in the South has displaced many quail coveys. Most of the vast native prairies that once provided excellent quail habitat were gone by the late '60s. As farming became bigger and more mechanized, fallow fencerows and field edges were bulldozed to make more croplands. These changes in land use killed many of the native plants that quail depend on. In their place, some less quail-friendly exotics were introduced. One of these is fescue grass for forage and hay. Fescue is literally murder for quail. Not only does it grow so tight the birds can't walk through it, but the plant is also extremely invasive and crowds out plants quail can use. Urban sprawl, the enemy of most wildlife, has also taken a great deal of former quail habitat.

Predation is another problem. In the old days, when most farms raised chickens, hawks and owls were shot on sight. I'm not suggesting that we return to doing this—it's illegal and ineffective. "If habitat can't support game in spite of predators, it simply isn't good habitat," said Aldo Leopold, the father of today's wildlife management. Other predators, however, such as opossums, raccoons, skunks, coyotes, and even foxes are on the rise, mostly due to the lack of trapping, and they all eat quail eggs, poults, and even, occasionally, adult birds. Controlling these critters within legal methods and seasons helps not only quail but also other ground-nesters such as wild turkeys. There has been some talk about predation of quail poults by wild turkeys, but it has been disproved. Turkeys are, however, extremely aggressive and will run deer off a food supply, so they may limit quail populations in some cases, although this has not been proven that I know of.

With decreased habitat, quail are more exposed to predation. They are also more susceptible to diseases and affected by bad weather conditions. Pen-raised birds may experience avian pox, blackhead disease, and quail bronchitis, and it's possible that some wild birds have disease problems as well.

In areas where waste agricultural grains are not available, annual food crops of small grains next to cover can provide important quail habitat. COURTESY TED ROSE

Providing Food for Quail

As with all wildlife, management for quail consists of providing, food, cover, and water. Quail are primarily seed eaters, although they eat some insects when they're readily available. Seeds of agricultural grains are very popular, with soybeans the number one choice and corn, milo, and weed seeds second. In some areas quail rely mostly on weed seeds, but they'll also take advantage of small grain crop residues (though fall plowing eliminates a lot of waste grain foods). The food portion of a quail management plan may include crop residues, native weed and grass seeds, shrub and tree fruits, and annual food plots.

Food plots. Food plots can provide excellent food sources. The plots must be big enough or numerous enough and also located close enough to cover that quail can get to them. The food plot size should be at least a quarter acre, and half-acre plots are even better. With this size plot you can plant a half or a fourth in grains and allow the rest to grow weeds for seed and cover. Ragweed grows the first year and provides great quail food. For many years I've planted food plots by simply broadcasting soybeans, milo, and millet, which is quick and easy. I break the ground well, apply lime, bring the soil up to test with fertilizer, and then broadcast the seed and go over it with a disk. Or you can drag a big cedar tree over the plot.

It's important that the plot be close to brush. A long, serpentine plot following a brushy fencerow is better than a square block. Islands of good cover within the plot are even better. Make sure livestock are fenced off the plot, as some grain sorghums can be poisonous to livestock after frost or drought. Plan to have at least one grain food plot for every forty acres of farmland, although if you leave crop residues, you can get by with fewer plots. An alternative is to leave rows of crops standing alongside cover and then mash half of them down in early winter.

Native foods. Providing native foods is another tactic for attracting quail. Important natural quail foods include:

- Acorns
- Asters
- Bedstraw
- Beggar ticks
- Blackberries
- Cinquefoil
- Clovers
- Crotons
- Dandelion
- Foxtail
- Goldenrod
- Grapes
- Korean & Kobe lespedeza
- Poison ivy
- Ragweed
- Sedges
- Smartweed
- Sunflowers

Providing Cover for Quail

Provide good cover throughout the quail's range. This means that you should have good cover interspersed throughout every forty acres. Fence off any natural cover such as brushy fencerows, ditches and waste or odd corners so livestock can't graze them and cover will grow. In agricultural areas, field borders or buffers are quite effective not only in providing food and cover, but also in erosion control and maintaining surface and groundwater quality. These borders can take on many forms, but essentially they are strips of land taken out of production and maintained in herbaceous cover. Their width may vary from as little as ten or twelve feet up to over one hundred feet, depending on your objectives. As a general rule, wider buffers will provide more cover and attract more birds. These herbaceous borders, through natural succession, will eventually be replaced by woody cover. They require maintenance—periodically disturbing the border—every two to three years. This can be a light disking or burning to maintain the area in weeds and to control woody plant succession. Mowing provides few wildlife benefits and in fact can be detrimental because it destroys the tall vegetation and creates a dense lower, horizontal layer of litter that eliminates cover and restricts quail movement. Don't do maintenance on

Specific types of cover for nesting, roosting, and travel are extremely important for quail. This cover not only protects them from predators, but from severe weather as well.

the buffers during the nesting season which, depending on the locale, runs from April 1st to September 1st.

Quail and other upland birds require three basic types of cover: escape cover, nesting cover, and roosting cover.

Escape cover. Escape cover should be available near all food sources. Good escape cover is often so thick a human can't walk through it. Brambles are an excellent example of good escape cover. Loosely constructed brush piles strategically placed near food sources can also provide escape cover.

Nesting cover. This type of cover is obviously important, but it's often unavailable. Nesting cover should be unmowed or ungrazed grasslands or field borders. The grasses should be redtop, timothy, orchard grass, perennial ryegrass, Kentucky bluegrass, or mixtures of native warm-season grasses.

Roosting cover. Quail roost on the ground out in open fields. Roosting cover should be clumpy but loose, with some overhead protection but not too much. Native warm-season grasses, fields of ragweed, and unmowed or hayed cool-season grasses such as orchard grass are good choices.

Overall Quail Management

Because of the tremendous drop in quail populations, the Northern Bobwhite Conservation Initiative (NBCI) was formed. NBCI encompasses all or part of twenty-two states from the East Coast to Texas, Oklahoma, and Nebraska. The initiative's approach (to provide food and cover on private land) is similar in some ways to that of the North American Waterfowl Management Plan but considerably more ambitious. The Southeast Quail Study Group estimates that habitat improvements are needed on eighty-one million acres to reach a quail population goal that equals 1980 levels. The overall goal of the NBCI is to restore and maintain bobwhite numbers to a total of 2.77 million coveys.

There is no "silver bullet" when it comes to restoring quail populations. Quick fixes, such as stocking pen-reared birds, making local habitat improvements, and passing restrictive hunting regulations, distract people from the important work of providing a wider-scale habitat. The NBCI has divided the eastern United States into fifteen bird conservation regions. Many private land programs, both state and federal, are available for quail restoration projects as well.

A common problem among those who manage land for quail is invasive cedars. Quail don't do well in cedars, and removing them not only allows for better food and cover, but it also eliminates perches for avian predators such as owls and hawks. If you have problem cedars, cut through some of them about knee high just enough to allow them to be pushed over. Although they'll fall, they'll remain alive for many years and provide escape cover when quail are fleeing coyotes or other ground-based predators.

When planning any habitat improvement for quail, keep in mind the need for connecting travel lanes or cover for quail to use when moving to or from feeding, roosting, or brooding areas.

Pheasants

Many of the management practices used to attract quail can also be used for pheasant management. Pheasant habitat requirements, however, do vary from quail somewhat. As always, food, water, and cover are the keys. Pheasants need heavy protective cover, including grass or similar cover for nesting. They also require weed patches and field edges for brood-rearing and loafing. Crop fields and annual weed seeds provide their food. Pheasants, as a general rule, are more tolerant of large areas of uniform cover than are quail. When landowners practice "clean" farming, food may be scarce, although it is usually available in sufficient supplies.

Providing Food and Winter Cover for Pheasants

The most critical components of pheasant habitat in most instances are winter food and shelter. Even in areas with sufficient available cropland foods, though, food plots can play a crucial role in winter survival, offering both food and shelter from the elements and predators.

Iowa DNR wildlife research biologist Todd Bogenschutz states, "There have been very few documented cases of pheasants actually starving to death in Iowa." Since starvation is not usually an issue, why are food plots so important for pheasants? The reason is twofold. First, according to Bogenschutz, food plots provide winter habitat as well as food. In fact, if properly designed and large enough, the habitat created by a food plot can be more beneficial to wildlife than the food itself. Second, food plots allow pheasants to obtain a meal quickly, thereby limiting their exposure to predators and maximizing their energy reserves.

"If hens have good fat reserves coming out of the winter, they are more likely to nest successfully," said Bogenschutz. He offers the following suggestions for planting food plots for pheasants:

Plant corn and sorghum. They provide the most reliable food sources throughout the winter as they resist lodging, or falling over and being covered, in heavy snows. Pheasants prefer corn to sorghum, although sorghum provides

Winter cover is often the biggest limiting factor in ring-necked pheasant habitat. Even agricultural areas will benefit from good annual food plots that provide cover plus food without expanding energy. Or leaving rows of sorghum grains standing alongside cover is also extremely effective winter cover.

better winter habitat. Sorghum is also less attractive to deer, a factor if deer may be a problem as competitors for food. Many states and Pheasants Forever chapters offer cost-share assistance for food plots.

Place food plots next to wetlands, CRP fields, and multi-row shrub-conifer shelterbelts that provide good winter habitat—but away from tall deciduous trees that provide a good place for raptors to sit and watch.

Base the size of the food plot on where it's located. If a plot is next to good winter cover, it can be fairly small—but a minimum of two acres. If winter cover is marginal, such as a ditch, then the plot must be larger (five to ten acres) to provide cover as well as food.

You may be able to leave food plots for two years. The weedy growth in the second year provides excellent nesting, brood-rearing, and winter habitat for pheasants and other upland wildlife. Food plots that experience heavy deer use generally need to be replanted every year.

An alternative to planting food plots is to leave several rows of standing crops near a shelterbelt or heavy cover.

Providing Additional Winter Cover for Pheasants

Pheasants typically move up to two miles from their summer range to where they overwinter. Once they arrive at the winter cover, they seldom travel more than a quarter mile to a food source, so easy access to food through travel

Other types of cover, including shelter-breaks, sloughs, and brushy travel lanes are also important pheasant habitats.

lanes is important. The quality of winter cover, such as wetlands, woody draws, old fields, and brushy hedgerows, is determined by vegetation density at ground level. Shrubby understory is best for protection from the elements. Here are some tips for providing additional winter cover:

Protect the natural cover. Weedy odd areas, brushy draws, stream banks, woody fencerows, wetlands, and marshes are used by pheasants for both nighttime roosting and daytime loafing. Creating a scattered pattern of preserved winter cover reduces the bird's susceptibility to predators. A ten-acre willow, cottonwood, or cattail slough can hold as many as a hundred pheasants, but since pheasants concentrated in a single wintering area during severe weather are more susceptible to predator pressure than those in less concentrated locations, winter cover areas should be no smaller than a quarter acre. A good rule is to maintain at least one to five percent of the hunting or wildlife managed properties in secure winter cover—and the more the better.

Provide short-term and long-term cover. Food plots and crops left standing provide short-term winter cover. Long-term or permanent plantings are most effective when they enhance existing brush or weed patches or lowland areas. These include windward plantings of low-growing and ground-hugging shrubs that create a closed canopy. The minimum width if the planting is next to existing cover is twelve feet. Winter cover should be located no farther than a quarter mile from winter food.

If you're enhancing existing cover, follow the cover. If you're creating new cover, use block plantings (i.e. plantings in shapes that are more square than rectangular or serpentine) that afford more cover and less edge, because block plantings don't fill with drifted snow as easily as long, narrow plantings do. Cover plots should be at least five acres, and bigger is better. The plantings should provide a closed canopy within five years. The best winter cover areas are multiple-row plantings—five to ten rows of shelter trees such as conifers. After the rows of conifers, continue with more strips of shorter, bushier types of trees and shrubs. In large shelterbelts, the windward outer row should be conifers. Then plant a strip of grass and finally plant a strip of shrubs.

The best cover plots have a travel lane that provides cover from one piece of permanent winter cover to another. These lanes should be a minimum of four feet wide and up to a maximum of twelve feet high. It's important to keep the shelterbelt maintained. Remove trees that extend above the shrub canopy or that have stout lateral branches to prevent hawks and owls from perching. Gaps in the plot can be filled with strong-stemmed weeds such as ragweed or sunflowers that resist flattening by snow.

Several years ago I hunted one of the best pheasant areas in the country, and I've never seen anything like it, including hunting the coveted Governor's Hunt in South Dakota. Located near the Kansas/Oklahoma border, the area consisted of vast acres of native warm-season grass in CRP broken into forty- to one hundred-acre fields. Bordering each field was a milo or sorghum strip at least one hundred yards wide. On each windward or north side was a shelterbelt cover area, and each of these was also bordered with a milo strip. Each drainage ditch was filled with native warm-season grasses, with plum thickets and old cottonwoods in the bottom. Every little marsh or swampy area had been expanded or managed to create heavy cover. Numerous stock ponds had been installed and fenced completely, providing many more marshes. The area held pheasants, pheasants, and more pheasants, waterfowl galore, and deer and other game.

Providing Nesting and Brooding Areas for Pheasants

Providing nesting and brooding areas is next in importance after providing cover. Kansas studies have shown that two of every three pheasant nests are located in wheat fields. Alfalfa and red clover hay fields combined with bromegrass or timothy are also excellent for nesting and brooding. Road right-of-ways and fencerows are also important. Don't mow or disturb grassy areas that may be nesting areas until after the last week of June.

Pheasant broods remain in close proximity to the nest site for the first three weeks after hatching, so nest cover is also brood cover. Delaying hay cutting and altering mowing patterns can reduce losses to hay field nests and broods. After the hay is cut, croplands provide cover. One of the most important management factors for pheasants is vast CRP fields, especially those with established native warm-season grasses. These areas offer large blocks of cover and food as well. Drought has the most serious effect on pheasant populations over much of their range because it causes weed and grass growth to be thin, resulting in less food and cover.

Mourning Doves

The most common method of habitat management for doves is planting fields to hunt. Standard agricultural practices must be used to follow the regulations against baiting. Black-oil sunflowers or the millets are preferred grains.

As mourning doves are migratory, most management practices involve providing food, cover, and water during the fall when the birds migrate and during the fall hunting season. Because of the nature of their feet, doves can't cling to upright stems or stalks, so seedheads on tall, upright plants are, for the most part, unattainable. The structure of their feet also prevents them from scratching, so seeds and foods covered by dense stubble or top growth are also unavailable. The best dove foods are seeds scattered on relatively bare ground. Wheat, corn, millet, weed seeds, or sunflower seeds scattered on bare ground with little in the way of stubble, stalks, and canes are preferred. It's important that the dove foods you provide be produced through normal agricultural practices that comply with federal dove-baiting regulations, and you must follow state baiting regulations as well. Deliberately scattering food for the purpose of attracting doves to hunt is illegal.

On row-crop acreages, certain farming practices can attract doves without a great loss in farm income. Early planting and combining of milo is one example. Doves also love wheat seeds, and combining the wheat seeds in July followed by clipping and baling the straw produces a very attractive feed field of waste grain, weed seeds, and bare ground. Early cutting of corn fields for silage and combining grain sorghums in late August will practically guarantee a good dove field.

Creating Dove Fields

Many landowners create feed fields (a feed field is usually several dozen acres) to attract doves. Before you do this, make sure you understand federal and state baiting and dove-hunting regulations.

Millet. Millets, including browntop and German, can be used to create dove fields. Seed it in widely spaced rows (thirty-six to forty-eight inches apart). Keep the strips well cultivated to prevent weed growth between the rows. These millets as well as the wild millets (including foxtail) that may grow in the rows shatter out, or lose their seeds, in late August and early September.

Black oil sunflower. This is one of the most effective seeds for dove fields. Prepare the seedbed as you would for growing corn. Apply lime, a balanced fertilizer with at least sixty pounds of nitrogen per acre, and a preemergent herbicide before planting. It's extremely important to plant the seed at the proper time so the seedheads ripen at the same time the doves arrive. Obtain planting dates from local conservation or country extension offices. In Missouri, sunflowers must be planted by mid-April in order to have mature seeds by September 1. Sow or drill the seeds one inch deep and twelve to fourteen inches apart in rows spaced about thirty inches apart. Cultivate once or twice before the sunflowers reach a foot in height. Do not irrigate—it encourages weeds and sunflowers are fairly drought-resistant.

Providing Roosting and Loafing Cover for Doves

Doves prefer food fields with nearby trees for roosting and loafing. Pine plantings, wood lots, and especially Osage orange hedgerows or even individual trees are preferred roosting and loafing sites. These same areas may also be used as nesting sites by local doves.

Providing Water for Doves

Doves usually drink early in the morning and late in the afternoon. Knowledgeable hunters know that farm ponds and stock tanks are great late-afternoon hunting spots. Doves don't like vegetation, even low vegetation, around their watering spots. The ponds they most often visit have bare ground around the water's edge. Ponds that are low in water, partially filled or even new and unfilled are perfect. Some landowners use draw-down pipes in dove watering ponds to create the ideal situation. If the pond is located near feed fields, so much the better.

PRAIRIE CHICKENS

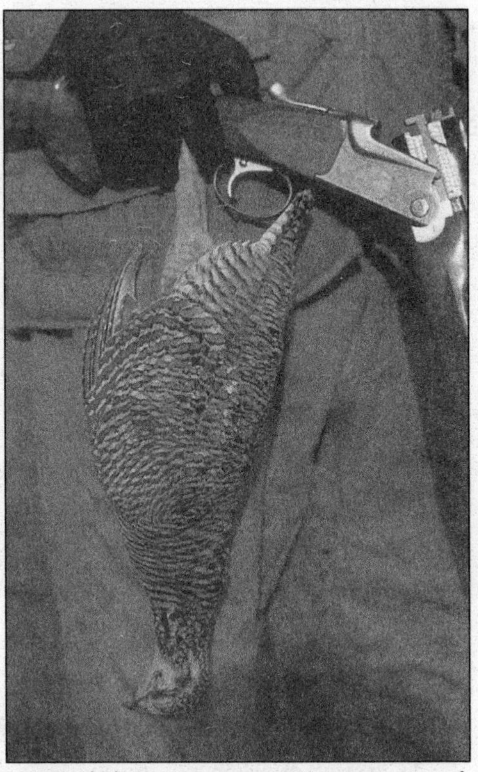

Prairie chicken management requires vast areas of native prairies. Recent studies have shown a mix of rotational grazing and burns can provide the diversity of the prairies once managed by buffalo grazing and natural or man-made fires.

Management for prairie chickens is basically the same as prairie management. These native grouse prefer open areas of good-quality permanent grass with only a minimum of brushy cover in draws or along fencerows. For nesting and roosting they prefer grass at least ten to twelve inches high and sturdy enough to withstand strong winds and inclement weather. Open hummocks with sparse, short grass cover are preferred for spring courting.

To attract prairie chickens, keep at least a third of the acreage within a prairie chicken range in permanent grasses. These can be taller domestic grasses, but native bluestem and other prairie grasses are preferable. One of the problems in managing a flock of prairie chickens is the vast acreage required. Prairie management, including prescribed burns, has been a traditional management method. Recent studies have shown that proper rotational grazing is also a major management factor. The grazing should emulate that of the herds of buffalo that once wandered the prairies. Prairie chickens require diversity in plant heights, cover and species. According to the Prairie Chicken Newsletter from the Missouri Department of Conservation, research has shown that grasslands that include bare or sparse patches interspersed with taller clumps of cover (or grasslands with a high degree of "spatial heterogeneity," or a great deal of variety) support more types and larger numbers of insects and birds. Mowing and burning do not achieve this ideal, but light grazing can. Research on the Konza Prairie in Kansas has shown that creating patches of burned, unburned, grazed, and ungrazed areas results in higher diversity of insects and more birds.

Ruffed grouse require a wide variety of woody vegetation. Management practices are used to keep woodlands in young hardwoods, regeneration stands, and mature stands along with other habitats such as old fields and brushy creek bottoms.

Ruffed Grouse

The ruffed grouse is North America's most widespread nonmigratory bird. Its range runs from Alaska to Georgia. Good ruffed grouse areas have a number of vegetation stages or types including young hardwoods, regeneration timber stands, field edges, brushy creek bottoms, abandoned homesites, and old fields. The best habitat has a very high number of woody plants or brush per acre. Habitat studies in central Missouri have shown that habitat that grouse use year-round has at least eight thousand stems per acre, and areas they prefer for drumming logs have nearly twelve thousand. Ruffed grouse prefer the brushy undergrowth stage of forest. Studies have also shown that fifteen percent of the year-round preferred food is furnished by high-canopy trees; forty-five percent comes from understory trees, vines and shrubs; and forty percent comes from grasses and forbs. Good grouse management consists primarily of creating interspersed plant communities to provide year-round foods.

Ruffed grouse management is timber management. Large, continuous stands of pole or mature timber or large conifer plantations do not support ruffed grouse. On larger tracts, the suggested management plan recommends an even-aged timber or clear-cutting program. An ideal program for both grouse and deer consists of five to twenty acres of clear-cuts scattered throughout the tract. The general guideline is to clear-cut about ten percent of your timber every ten years. On smaller properties, try group selection cuts, creating smaller openings, or miniature clear-cuts. Even a two- to three-acre opening can create good grouse habitat. Standard TSI methods don't produce enough of a clearing for grouse, as the canopy closes too quickly. Clear-cutting strips from sixty to ninety feet wide along field edges, roads, and firebreaks can also provide habitat. Regrowth of the brush will provide grouse habitat for up to twenty years. Creating a field border is also an excellent practice. These areas must be fenced to eliminate cattle grazing.

Aspen stands are preferred grouse habitats in the northern woodlands and are also used by woodcocks. As with hardwood stands, the more diversity of plant ages the better. Three age classes are required for good grouse habitat. Sapling stands of four to fifteen years are used for brood cover. Small pole-stage stands from six to twenty-five years provide fall and spring cover. The older aspens—forty to sixty years old—provide food as well as nesting and winter cover. These should all be well interspersed over the territory. There should be plenty of food sources, breeding and winter cover within the six- to ten-acre territory each breeding male claims.

Logging roads, trails, and other open lanes can be planted in white Dutch clover, a favorite food of ruffed grouse. Lime, fertilize, and seed as for other clover food plots.

CHAPTER 29

SMALL GAME

Small game animals—rabbits and squirrels—are favorites of many hunters. Habitat management for small game is fairly simple, and many of the practices recommended for other species can improve small game habitat as well. Some practices can also be developed for small game that will improve habitat for other wildlife, both game and non-game.

RABBITS

Given the fact that the average home range of a cottontail rabbit is less than five acres, it's fairly easy to establish good rabbit habitat even on small acreages. The requirements for good rabbit habitat include lots of well-distributed cover, a safe place for nesting and rearing young, and a plentiful year-round food supply.

PROVIDING COVER FOR RABBITS

The most important feature of rabbit habitat is well-distributed cover. Fence off woodlots so they're not grazed. This lets grasses, shrubs, and forbs come in naturally, providing both cover and food.

Rabbits live in relatively small areas and don't have many special requirements. Areas of quality grasses, legumes, and forages provide nesting, food, and cover.

Fencerows. Overgrown fencerows are wildlife havens. Wildlife such as quail and rabbits can move from them into bordering fields and quickly return to them for safety. Fencerows can also provide winter food, and this is extremely important when deep snow covers the ground, because rabbits can feed and find shelter without leaving the fencerow. One of the easiest ways of creating better fencerow habitat for rabbits and quail is to top tall trees. With more sunlight, shrubs and grasses do better and growth is kept low and dense. If the trees are cedars, cut them halfway through and push them over to create living brush piles. Other trees, such as hedge and the locusts, will also remain alive for several years when you lop them over leaving the trunk partially attached to the stump.

"Odd" areas. Fence off gullies, ditches, ponds, dams, and other miscellaneous areas so cattle can't graze them. This prevents erosion problems caused when animals tramp the soil and graze soil-holding vegetation, and it also improves the area for wildlife such as rabbits. Allow these areas to grow into briars, brush, and sprouts. Top the sprouts when they become big enough to shade grass and shrubs, and keep part of the areas in grass to provide nesting areas. Keep pond dams free of woody vegetation to prevent roots from growing into the pond and causing leaks.

Brush piles. Brush piles are rabbit "condos." When you bulldoze to clear areas, don't burn the dozer piles. Push them to the edges of fields and into gullies and other odd areas. They provide instant shelter, and native plants usually grow up through their tangles to provide feed. In fact, rabbits will move in the first night a brush pile is constructed!

You can also create brush piles without a dozer. The best are open at the bottom so rabbits can move about easily. To create a pile like this, first build a "foundation" to hold the pile. It should be something that won't quickly rot and allow the pile to fall in on itself—rocks, discarded farm equipment, old culvert pipes, even logs. Old tires can be used, but cut them in half so water can't collect in them and create mosquito-breeding areas. Then start piling on the brush, beginning with larger limbs and adding smaller limbs and brush as you reach the top. The pile should be at least sixteen feet in diameter and four to six feet high.

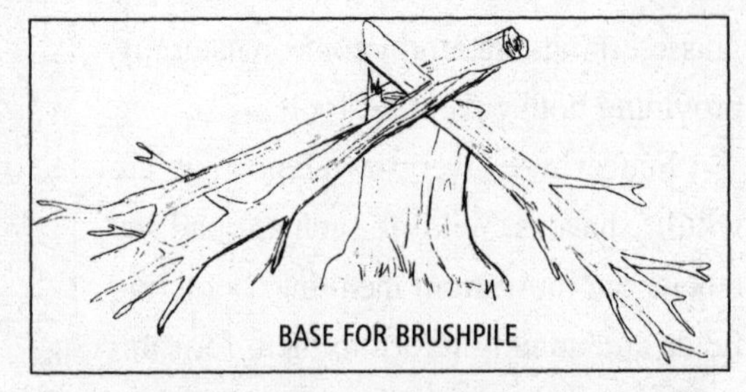

One of the best habitat management practices for cottontails is building rabbit "condos" or brush piles. They should be well supported on the bottom to provide runways and den areas.

Place brush piles close to other cover, travel lanes, and food sources. Rabbits won't use a pile built in the middle of a heavily grazed field. Good places to build them are near woods edges, fencerows, unpastured grasslands, and odd areas overgrown with briars. The number of brush piles required depends on the area, but in most cases, the more the better.

Providing Nesting Areas for Rabbits

Rabbits need protected grassy locations for nesting. They'll nest in the grass of orchards, pond dams, and ditch banks—even in your lawn. Terraces are also excellent nest sites.

Provide good drainage. The most important factor to consider in creating a nesting area is good drainage. Low-lying and flat areas are problematic during rainy periods, as the young rabbits are either drowned or forced from the nest. In flat areas, try creating a mini terrace by plowing two furrows against each other and seeding the area with a grass-legume mixture such as orchard grass and clover.

Protect nesting sites from grazing, mowing and burning.

Grow native warm-season grasses. Native warm-season grasses probably provide the best nesting situation. These plants have tall, stiff, upright stems with leaves on the upper portions. They reduce wind speed, modify transpiration and humidity extremes, and soften the impact of raindrops. The sturdy upright stems persist through the winter months and provide shelter during snow and ice storms. Growing in clumps, these grasses provide natural travel areas between the clumps and under the overhead cover they create. Young rabbits climb into the clumps for protection from the elements. The clump growth also allows room for germination of broad-leafed plants, providing food as well.

Providing Food for Rabbits

Rabbits will eat a wide variety of plant foods, but they can be choosey, and they eat some plants only during certain times of the year. You may wish to provide food plots for them.

Favorite foods. Bluegrass is one of the rabbit's most preferred plants except during the summer months when clover is the preferred food. After bluegrass, clovers—especially white clover—Korean and Kobe lespedeza, crabgrasses, timothy, orchard grass, knotweed, and wheat are preferred. During fall and winter, wheat as forage and grains such as corn, soybeans, and milo are important foods. as are white clover, bluegrass, timothy, cheat grass, and orchard grass.

Clovers and lespedeza interplanted in native warm-season grasses provide a good food source.

Food plots. Food plots can be created for rabbits. It's best to have several small plots located next to a diversity of cover sites rather than one large plot. Deer love the same foods rabbits do, however, and they can be a problem if you grow small plots. The best plots are planted in strips. A plot twenty feet wide by two hundred feet long is a good general size. Place the plots close to good cover such as brushy draws, fenced woodlots, fenced ponds, brushy fencerows, or brush piles.

The best time to prepare a rabbit food plot is in August or early September. As with other food plots, take a soil test first, then lime and apply fertilizer according to the test results, working in the fertilizer when you prepare the seedbed. After you've created a smooth seedbed, broadcast a half bushel of wheat per acre along with five pounds per acre of inoculated ladino clover, alfalfa, red clover, Dutch white clover or hairy vetch, or two pounds per acre of bird's-foot trefoil.

The wheat will die out the first year, but the legumes will provide browse for three to five years. Clip the plot twice each year (around June 10 and September 1). Although the plot should be fenced and protected from burning and excessive grazing, light grazing to remove about half the growth in mid-June is even better than clipping. Every other year, top dressing with one hundred pounds of potash and one hundred pounds of phosphate per acre will keep the plot producing. In thick, heavy stands of grass, rotational grazing can also be used as a management tool.

The main ingredient of prime squirrel habitat is woodlands, woodlots, fence rows, or other areas with mast-producing trees, along with soft mast and, in some instances, waste small grains.

SQUIRRELS

Producing good squirrel habitat can be easy or quite tough. It can take more work and time than creating rabbit habitat if you don't have the right materials on hand. For the most part the "right materials" are trees. Any woodland having from fifty to seventy-five hard mast-producing trees, including oaks, hickories, walnut, elms, and maples, along with soft mast trees such as mulberries, will produce squirrels. This can include small woodlots or even fencerows with trees. You may be able to enhance the squirrel carrying capacity of many types of woodland with improved food and cover.

Providing Food for Squirrels

Squirrels are omnivorous—they eat both plant and animal matter—although their main diet consists of plants and their main food is acorns. Bumper crops of acorns or other mast produces bumper crops of squirrels. Likewise, a mast shortage one year leads to smaller numbers of squirrels the following year.

Other foods squirrels eat. Other foods are also important during different times of the year. In late winter and early spring, about the time the first litters arrive, squirrels feed on the flowers and buds of hardwoods, including maples, elms, oaks, and sweet gums. In late spring they switch to mulberries and mushrooms. By summertime, they feed on bramble berries, wild cherries, wild strawberries, the leafy portions of herbaceous plants, corn in the milk stage, and orchard fruits. In early fall they turn to wild plums and wild grapes. By mid fall they begin "cutting" nuts (opening them with their teeth to get to the nut meats). This begins with the first nuts to ripen, the hickories. The squirrels progress through the other nuts as they ripen: hazelnuts, walnuts, beechnuts, pecans, and acorns. Other fall foods include corn, pokeberries, cypress, tupelo gum, pine seeds, and the fruits of the honey locust and sugar maple. Squirrels bury acorns and other nuts as caches to carry them through the winter months.

Feeding stations. In harsh winters with low mast crops, you can help squirrels survive by providing feeding stations with whole ear corn. These stations can also be valuable in early spring before many trees have budded out. You should supply about three pounds of unshelled corn per squirrel per week. Wire baskets or hoppers fastened to tree trunks make simple feeding stations. One station per twenty-five acres is usually sufficient. Make sure to provide feed regularly once you start.

TSI. Solid stands of timber can often be improved for squirrels through TSI. Selective harvesting of a few trees opens the canopy, allows for better mast production, encourages understory growth, and creates an uneven-age stand. This ensures a more dependable food source and also allows for the natural creation of den trees. Leave some less-timber-valuable trees for squirrels. Remember that the acorns mature at different times: white oaks mature in one year, black oaks require two growing seasons. If a timber stand has evenly divided amounts of white and black oaks, an average of one elm, one maple and one mulberry, plus about half a dozen hickories, it will provide plenty of staple food.

Providing Cover for Squirrels

Squirrel cover consists of trees, small woodlots, large tracts of timberland, and hedgerows or fencerows with trees.

Preferred cover. Fox squirrels prefer mature hedgerows and small woodlots with canopy openings as well as park-like pastures with scattered trees. Lightly grazed woodlots are, in fact, well used by fox squirrels. Gray squirrels prefer larger, more densely wooded areas. They like closed canopies with a good amount of ground cover of herbaceous plants, saplings, and shrubs. Downed treetops left in the woods are also attractive to grays.

Den trees. Squirrels nest in loose leaf nests during summer months or even year-round in the absence of tree cavities. One of the most effective management practices is creating or maintaining den trees. These have either natural or created cavities that are used by squirrels and other wildlife including opossums, raccoons, and numerous birds. Natural cavities may require years to develop and are typically formed as the lower limbs in a forest stand are shaded, die, and drop off.

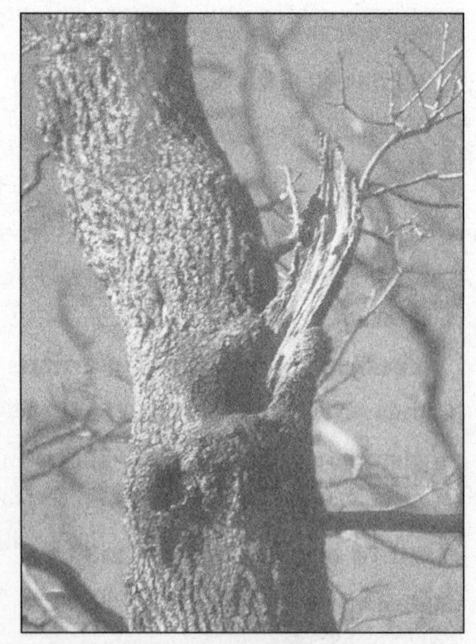

Leaving or creating den trees, especially wolf trees, is one of the best squirrel habitat management practices.

You can create den trees by girdling and/or applying herbicide to kill less valuable mature trees. Wolf trees—those with large, spreading crowns—are commonly removed in forest protection management practices.

The best use for them, however, is as den trees. Not only do the dead trees provide lots of den cavities, but killing them also creates an opening in the forest canopy. Leave one snag or den tree larger than twenty-inch DBH, four snags ten- to twenty- inch DBH, and two snags six- to ten-inch DBH per acre, but don't kill trees to create snags or den trees in limited-timber areas such as along streams, in small woodlots, or in fencerows.

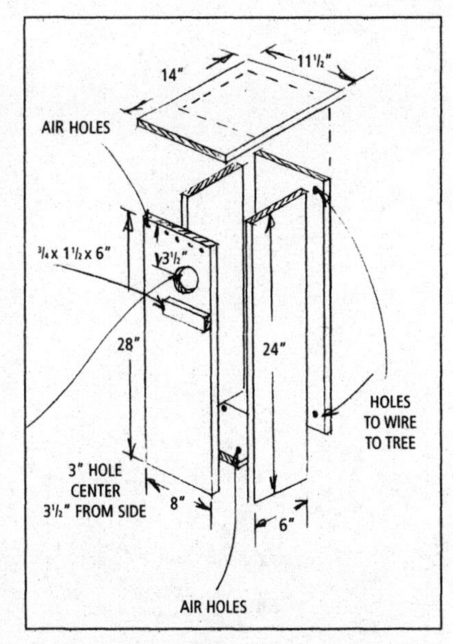

In areas with insufficient trees to produce dens, artificial dens can be created and fastened to trees.

Nesting boxes. You can create squirrel boxes, or artificial dens, and fasten them to trees. Allow three dens for each squirrel family (which consists of mother and young; males seldom use nest boxes). Place the dens, one to a tree, several yards apart and about twenty feet off the ground.

Providing Water for Squirrels

Finding enough water is usually not a problem for squirrels, although they tend to live near water given the opportunity. Small wildlife watering ponds, one per forty acres of timberland, can provide water for squirrels and other wildlife.

Planting More Trees for Squirrels

Planting hardwoods to create squirrel habitat is a life-long project. It takes thirty to forty years for many oaks to grow enough to create mast. There are, however, trees and shrubs that produce mast more quickly, including sawtooth oaks, mulberries, redbuds, hawthorns, and dogwoods. Allowing openings to grow into brambles and allowing wild grapes to grow on selected trees can also create food sources. In fact, most of the practices described in Chapter 16, Soft Mast Management, can increase squirrel food sources. These, as well as hardwood mast producers, can also be planted in fencerows, odd field corners, and other areas to provide food and cover.

CHAPTER 30

WATERFOWL

Waterfowl habitat management consists of two practices: creating nesting habitat and creating migratory food and cover habitat. The first, of course, is practiced primarily in the northern nesting areas of North America, except for wood duck nesting boxes in the South. Creating migratory food and cover habitat for both ducks and geese can be practiced on any portion of the migration routes.

Waterfowl habitat consists of nesting and migration habitat. The first step in habitat management in most instances is preserving potholes, marshes, and sloughs or creating new marshes, ponds, lakes, or flooded timber or fields.

CREATING NESTING HABITAT

Most ducks and geese nest in potholes, sloughs, and marshy areas, and the preservation, expansion, or creation of these areas is the key to nesting habitat management. The best nesting ponds have more than an acre of open water with scattered emergent plants such as bulrushes and cattails.

WATERFOWL NESTING PREFERENCES

Puddle ducks—such as mallards, pintails, shovelers, teal and gadwalls—and geese nest on dry land. They choose grainfields, hay fields, and other grassy, unused areas. Nests are usually within one hundred yards of water, but some may be as much as a mile away. After hatching her eggs, the female takes her brood to the water. The diving ducks, including scaup, canvasbacks, redheads, goldeneyes, and ringnecks, nest among emergent plants or along the shoreline. A general goal is for each acre of water to produce two to four young ducks per year. If your potholes are doing that, you probably don't need improvement.

Improving Water Supplies

Drought, resulting in potholes drying up in the summer months, is the most serious problem in waterfowl production. To provide more water:

Deepen a pond. Increase the depth of a pond or pothole by digging or blasting pits in a marsh. Pits should be two thousand to five thousand feet square with one for every one to five acres of marsh. A crescent- or L-shaped pit works best. One side of the pit should have a slope of no more than five-to-one (the slope drops one foot below the water surface for each five feet it extends into the water). One fourth of the pit should be no more than 3½ feet deep. Use the soil from digging to create an island for loafing and nesting. It should be twenty-five feet or more in diameter and have a settled height no less than two feet above the water level. A portion of the island should also have a five-to-one slope. Smooth the surface and plant it with grass and legumes immediately after you create it to prevent erosion and wind problems and to provide nesting cover.

Control plants. A common problem in marshes and potholes is overgrowth of emergent plants like cattails. Several methods can control these plants:

- *Apply herbicide.* Spray herbicide in strips forty feet or so wide across the area from the shore to open water, using only products that are safe for this practice. Check with your local SCS office or state wildlife agency about appropriate products.
- *Mow.* If the marsh is dry, you can control the plants by mowing strips. Mow down to ground level before viable seed is formed, usually in late June to early July, and mow again when regrowth is about two feet high. Don't mow or treat everything, just the strips.
- *Invite muskrats.* Muskrats will help control the plants. Use a scythe or herbicide to create open water and place hay bales in the opening. Muskrats will use the hay bales as places to build their houses, and they'll continue to open the water as they eat the cattails. Ducks will then use the houses as loafing places.

Grow more plants. Although it's a good idea to control cattails, potholes with bare shores are not attractive to ducks, and grazing and trampling of the shore by cattle is a common problem. It's a good idea to keep a well-sodded area at least forty feet wide around a pothole or marsh. Fence the pothole and the area to keep the cattle out. Eliminate trees from the site to prevent avian predators from preying on the eggs and young ducks.

Create potholes and ponds for ducks. You can create structures with or without water-level controls. A structure without water level controls is simply a dam blocking a natural channel, and this is the least expensive to build. A structure with water-level control is more expensive, but being able to raise or lower the water level allows easier management of the pothole or pond. Before you build any water structure, check state and local regulations and contact your local SCS office. They can help design the structure and suggest management practices.

Potholes, ponds, or marshes with water level control devices are the most effective as the water level can be used to manage the vegetation.

The best sites allow fifty to seventy-five percent of the land to be flooded to a depth of about four feet. This usually provides plenty of open water without emergent plants taking over. If the pond is from two to four acres, build a small island in the center of it. Potholes may be no bigger than half an acre; ponds are usually larger. Potholes are usually natural; ponds are most often man-made.

Water level control ponds or potholes can be constructed in the same manner —with a dam across a natural channel—or they can be built without natural water inflow if a dependable water source is nearby. For instance, water can be pumped or diverted from a nearby stream. Better yet, construct a storage pond above the pothole and then release water from it when you need to flood the pothole pond. Grow duck-food plants in the lower pond during the summer while the upper pond is producing young ducks.

CREATING MIGRATORY FOOD & COVER HABITAT

Across the flyways, many public and private properties use a variety of practices to attract and hold ducks and geese. These practices include flooded timberlands, flooded grain fields, farm ponds, backwater river sloughs, and reservoirs.

The key is providing both food and water. Ducks and geese will readily feed on dry land. Ducks, however, prefer food that is underwater, and ducks are heavy eaters. Each mallard eats from one to two pounds of grain each week. A duck pond or marsh can hold ducks, and flooded timber bottomlands are mallard havens. A duck field, however, is the most attractive to ducks.

CREATING PONDS AND RESERVOIRS

The construction of a pond or reservoir is basically the same as described for nesting potholes. Those with water-level controls and dependable sources of water are the best. Most ponds, which are usually stock-watering ponds, are not attractive to ducks unless they are designed somewhat differently.

Fence off the pond to allow vegetation to grow around the shoreline. A frost-proof stock watering tank below the pond dam can provide water for livestock. Most livestock and fishing ponds are fairly deep throughout with little shallow water.

Make sure at least one fourth of the surface area is two to three feet deep to allow for a small "marsh," but be sure to control cattails and emergent plants.

Add a grassy island in the center to make the pond more attractive to ducks.

The double pond system mentioned earlier can also be used to grow foods for fall ducks.

FLOODING BOTTOM TIMBERLANDS

Diking bottomlands and flooding them in the fall is a tradition in the South, and these "green-timber" areas are mallard havens. The ideal depth is one to fifteen inches. A water level control is needed, and the water can be pumped from a dependable source into the bottomlands. Creating small openings in the timber and planting with Japanese millet or smartweed makes the area even more attractive. In most instances the control gates are closed in October and the area is flooded and is left flooded until the first of March. Winter flooding of bottomlands increases the growth of many hardwood trees. The area, however,

Flooding bottomland in the fall is an excellent tactic for attracting migrating waterfowl. The areas should be left flooded until late winter to provide food before the ducks head back north.

must be drained and allowed to stay dry through late spring and summer because year-round flooding kills the trees.

Flooding rice fields also offers lots of food and rest areas for waterfowl.

Flooding rice fields. Flooding rice fields in the fall and winter is also a tradition in the South, one that can provide as much as one hundred to two hundred pounds of food per acre. Disking down an acre or two of the stubble makes the field more attractive to ducks.

Flooding marshes. Natural marshes can also provide food and cover during migration. They require the same maintenance as described for potholes. Draining in the summer, managing the vegetation, and flooding in the fall is the best tactic. A large number of natural plants which are eaten by ducks are found in most marshes. Some plants, however, are not good duck foods. It's important to know the proper plants and how to manage them. Giant cutgrass, reeds, and maiden cane will usually dominate marshes unless they're controlled. Ducks don't eat them, but cattle do. Cattle, however, do not prefer the good duck foods. One management tactic is to graze the marsh during the summer.

Flooding marshes, sloughs, and other low-water areas provides prime habitat. In most instances the areas are managed to produce natural plants waterfowl prefer.

Over time, dead stems and leaves build up in a marsh. This eliminates open water and makes feeding difficult for ducks. Burning removes this debris. Burning can also reduce the coarse perennial plants and create a better environment for growing natural duck foods. The marsh should be burned at least every other year in the winter or early spring months.

CREATING DUCK FIELDS

You really can't beat a duck field for attracting ducks. A duck field is basically a low-lying area that can be drained in the spring, planted in field crops, and then flooded during the fall and winter to make the crops available to ducks. Duck fields can be created below a farm storage pond, reservoir, or near other dependable water sources. The key is to have the water available at the right time. The amount of water available controls the size of the field. You must construct a dike around the field and install a water-level control in the dike. The water level should be no more than fifteen inches for puddle ducks that tip up to feed. If the field is sloped, one tactic is to gradually flood so a steady source of food is available. Keep the field flooded until early April so ducks have plenty of food for their northward migration.

The field can be planted in grain crops or "duck foods" and milo and corn are some of the most popular. Corn is treated the same way as field corn, with liming, fertilizing, and planting in a well-prepared seedbed. One of the most effective places I've hunted consists of forty acres of corn with a permanent blind in the center. A center portion in front of the blind is harvested to create

Duck fields are hard to beat. They are diked fields that are planted in prime duck foods in the spring, then flooded in early fall. Foods range from corn to millets.

an opening. Corn does require more water to flood, but it continues to fall over throughout the winter, providing a continuous supply of grain. Ducks will eat corn in the water, under the water and hanging a foot above the water. Check state and federal baiting laws. It is illegal to knock down or drag, shuck, or shell the corn, as this is considered baiting.

Duck Foods

Some of the best foods for both duck fields and marshes where the water level can be controlled are Japanese millet, browntop millet and smartweed. These and special "duck food" packages are available from several seed dealers, including Pennington and Mossy Oaks BioLogic products. Pennington's Duck Mix contains Japanese millet, white proso millet, buckwheat, and Penngrain DR grain sorghum.

MILLETS

The millets are annuals, and the seed crop matures about sixty days after planting. Fertilized properly, it will yield about 1,500 pounds of seed per acre. Prepare a good seedbed in early July. Drill seeds about a half inch to one inch deep or broadcast and cover them with a disk or drag. Apply at least five hundred pounds of 5-10-10 fertilizer per acre. As soon as the plants mature, flood the field.

SMARTWEEDS

Smartweeds are a natural duck food, but the seed isn't readily available. Smartweed seeds are present naturally in many wet areas. Disturbing the soil slightly will help the seed to germinate and establish a stand.

ATTRACTING GEESE

Management of goose habitat is a little different from management for duck habitat, but it's simple. Geese prefer large, open reservoirs for spending the night and loafing during the day. They feed primarily on waste grain in agricultural fields and green wheat fields. They also love the lush growth found

Attracting and holding geese is easy if you have large agricultural fields with nearby reservoirs or large lakes for resting.

Wood duck nest boxes are easy to build and quickly attract these local ducks of the South and Southeast.

on many lakeside lawns, golf courses, and city parks. Many geese have become "local" and do not migrate. This is an increasing problem. If you wish to attract more geese, however, offer them nesting tubs, but be aware that they can become a nuisance.

Attracting Wood Ducks

One of the most successful waterfowl management practices is the use of wood duck nest boxes. Wood ducks nest in hollow tree cavities, usually within a mile of water, but preferably nearer. Nest boxes are easy to build and can be used to attract and raise the beautiful little birds. It's important to use predator guards on the supports for the boxes to keep raccoons, opossums, and snakes from raiding the nests and eating the eggs or young.

CHAPTER 31

RIPARIAN CORRIDORS, SPRINGS, AND SPRING SEEPS

Providing a permanent water source is a major goal of wildlife habitat management. Most wildlife needs a steady and ready supply of water. Some animals get their water from skimpy sources such as dew; others require a more abundant source. Water can be in the form of natural lakes, created ponds and reservoirs, creeks, rivers, oxbows and sloughs off rivers, marshes and wetlands, springs and spring seeps (areas where water seeps out of the ground, but does not run off)—even such unusual sources as strip-mine

A permanent source of water is invaluable for most wildlife habitat management practices. This can mean natural lakes, ponds, reservoirs, creeks, rivers, oxbows, sloughs, springs, and even stock tanks and mining pits.

pits. The previous chapter covered preserving and creating wetlands for waterfowl; this section discusses preserving and creating riparian or stream corridors, spring seeps, ponds, stock tanks, and reservoirs.

Preserving Riparian Corridors

Some of the most important water sources for both wildlife and mankind are streams. In the drier, western portions of North America, riparian zones (areas along the banks of natural waterways or lakes) are extremely critical to wildlife survival. Protection of riparian corridors is a major factor in providing good water sources for wildlife and quality water for human use. Riparian corridors may consist of grasslands alone; however, in most instances, even in dry country, they consist of woodlands. A good number of wildlife use these areas for all or a part of their habitat, and some animals spend their entire lives in these zones. Riparian zones provide a steady supply of food, cover, nesting places, dens, and wildlife travel lanes. To preserve riparian zones:

Fence the area off from livestock. This protects the area, including wooded areas, from livestock damage such as vegetation destruction and erosion caused by denuding and destruction of the banks. Develop alternative watering

The first step in all instances is to fence livestock away from the water source. On riparian corridors a one hundred- to two hundred-foot-wide strip should be planted in trees, shrubs and/or grasses and protected from livestock as well.

sources for livestock; funding programs and help may be available for creating them. Adding alternative sources may be as simple as creating an area of limited access to the stream for wildlife; more sophisticated systems may use solar pumps to pump water to a stock tank.

Practice TSI. At least a one hundred- to two hundred-foot-wide corridor of trees should be protected or planted to protect the zone. Because these bottomland areas have excellent soil, they often produce valuable hardwoods that can be managed for both monetary gain and wildlife habitat using standard woodland management practices such as TSI. It's important not to overcut or clear-cut any areas of a riparian corridor, and timber cuts should minimize the loss of important snag and den trees. When you make a timber cut, don't remove only the tall trees such as walnuts, cottonwoods, and sycamores. Cutting out the tall trees encourages understory trees such as mulberry and redbud to grow more thickly, preventing the growth of the new tall-tree seedlings. Even if the zone is not as wide as one hundred to two hundred feet, fencing it off and allowing natural succession will result first in woody shrubs and vines, followed by sapling growth, and finally timber growth. The faster-growing trees in the South and Southeast include walnuts, cottonwoods, green ashes, willows, silver maples, sycamores, elms, sweet gums, and yellow poplars.

Protect stream banks. It's important not only to protect vegetation along the stream but to protect stream banks themselves.

- *Don't channelize the stream.* (This was a common practice several years ago that involved straightening a stream to create a channel to move water faster).
- *Don't remove trees that have fallen over into the stream or those that appear ready to fall.* The fallen trees help slow the current, hold the bank in place, and also create important wildlife habitat.
- *Don't create or enlarge stream crossings except where necessary,* and then choose areas where you'll do the least damage, such as natural low-water crossing areas. Areas with high banks that must be cut will wash out each time there's high water.
- *Make sure that crossings are at right angles to the stream.*
- *Stabilize the stream bank if necessary to prevent further erosion problems.* One way to do this is to simply cut switches from willows and stick them in the bank. A good number will sprout and create a

natural bank cover. Sometimes a revetment (a facing of concrete or stone) may be needed and this may consist of several practices. One practice is using rocks to create riprap to prevent erosion. A second method is to plant willows. Special revetment bars are also available, although this is more costly. Funding is usually available for this type of stream protection.

Preserving Springs and Spring Seeps

In many parts of the country springs, or spring seeps (places where water oozes to the surface and forms a pool) are very important wildlife water sources. As the water is not groundwater, it stays in the fifty to sixty degrees F. temperature range even during freezing weather. This makes these areas especially valuable during the winter months. The year-round moisture also encourages growth of a wide variety of green vegetation that wildlife such as turkeys relish and need during the critical winter months. To preserve springs and seeps:

Springs are an important water source in many parts of the country. Again, they should be protected from livestock. Sometimes overhanging trees should be removed to provide more growth of grasses and forbs surrounding the seep or spring.

Fence them off if possible to exclude livestock. The animals will trample them and can create muddy water and disease problems. (Appropriate livestock watering sources can be created downstream of the seep.)

Protect them from logging or other heavy equipment damage.

Practice TSI if necessary to create an opening around a seep. If mature oak trees surround it, there is usually not a problem. Shorter understory trees can, however, cut off sunlight, allowing little browse and few forbs and grasses to grow around the seep. Judicious thinning of these trees can provide a better wildlife habitat around the seep.

Before you alter springs or seeps, check to see if it's legal. Digging or enlarging of springs or seeps falls under the regulations of the Army Corps of Engineers and your state. Make sure you check with USDA Extension Service agents, the Natural Resources Conservation Service (NRCS), or local and state foresters before altering springs and seeps.

CHAPTER 32

Ponds, Stock Tanks, and Reservoirs

Every landowner dreams of a pond or lake for fishing and family fun. Ponds can be small, simple wildlife watering holes, stock watering ponds, fishing ponds, or a combination of the three. Lakes and reservoirs are simply larger impoundments. Regardless of size, in many instances creation of ponds is regulated by state and possibly federal regulations. Constructing a good, safe pond, lake, or reservoir requires proper location, design, and construction. Design help and information are available from the Soil Conservation Service (SCS), Natural Resources Conservation Service (NRCS), and in many cases from state wildlife agencies. The help of these agencies will be invaluable in creating your pond.

Ponds are one of the first projects with many new landowners. Ponds can be created for fishing and family recreation, livestock watering,1 and as a watering source and habitat for a wide variety of wildlife.

Ponds may be cold-water, cool-water, or warm-water. The difference depends on the water source and your goals. It's extremely important to assess your property to determine possible pond sites and the condition of the drainage area; to decide what type and size of pond or reservoir you want to construct; and to decide what you want to use the pond for. Ponds may be constructed to provide water for livestock, household recreation, irrigation, fire protection, fish production, or wildlife or a combination of several of these uses. This book is concerned with ponds for wildlife use.

Ponds for Wildlife

On sites without a permanent water supply, wildlife ponds can enhance the suitability of the site for wildlife habitat. These ponds are typically small—a quarter acre or less. I have one located below a spring seep that's only twenty feet across. Water stays in it year-round, even during droughts. It's regularly visited by all types of wildlife including turkeys, deer and quail. Ponds of less than half an acre are usually too small to successfully raise fish, but they can be used to raise catfish if you feed the fish. In areas without a permanent water source, one wildlife pond for every forty acres will provide water for many wildlife species.

Ponds should also be fenced to keep livestock out of the pond and surrounding buffer zone. Alternate livestock watering sources can be developed. The frost-free waterer shown is located below a pond on the author's property.

Pond Types

Embankment ponds. These are created in areas with a depression or where the ground slopes from gently to moderately steep, and surface water is used to fill them. A dam is constructed across the depression, and it impounds water above the original ground surface. This is the most common type of pond. The key to building an embankment pond is having the right topography to create an economical dam.

Excavated ponds. These are also called "dugout" ponds. A large portion of their water is stored below the original ground surface. Excavated ponds are constructed in flat areas where there is no depression or channel to dam. They're much more expensive to construct than embankment ponds because they require more earth moving. They may be filled with water seeping from under the ground or from surface runoff.

Combination ponds. Many ponds are created by using a combination of embankment and excavation.

Pond Tips

Ideal locations. The topography of the land will determine how much water can be stored and the amount of earth that must be moved to construct the pond. Economically speaking, the best site will store a large amount of water with the least amount of earth moving. The ideal location is one where a dam can be built across a narrow drainageway adjacent to a large area in which water can back up.

Water supply. Regardless of which type you build, you must have a reliable water supply such as a spring above the pond, surface runoff, ground seep, or a combination of these. It is extremely important to avoid polluted water sources. Don't locate ponds below feedlots, sewage lines, dumps, mine tailings, or other such sources of pollution. It may be necessary to create a channel to divert pollutants away from a pond site. Ideally, you'll own all the land that drains into the pond; this allows you to control pollution and other problems. If a neighbor's land in the pond drainage area has erosion problems, one method of control is to build a debris-collecting basin above the pond.

Soil type. The type of soil in the area is also important. It must hold water, and this usually means that you must have clay-based soil rather than sand-based soil. If you don't know what type of soil you have, make soil borings to determine the suitability of the site.

Drainage area. The condition of the drainage area is also important. If it is not well vegetated, erosion will create sediment in the pond, which can reduce the pond's holding capacity and create a muddy pond. Grassy (not over-grazed) fields or ungrazed (protected from livestock) woodlands work best. The size of the drainage area is an important consideration in construction of the pond and pond dam.

Small wildlife watering ponds spaced one for each forty acres of woodlands or where any other type of permanent water source doesn't exist, can make the area more attractive to wildlife.

Help. Get the help of experts as you determine the design and location of the pond, dam, and drainage area.

Avoid hazards. Make sure the pond is not constructed where buried cables or pipelines are present. And avoid building under overhead power lines, as they can present a hazard.

Ponds for Livestock, Fishing, and Wildlife

Most folks these days build ponds for fishing, water recreation, and wildlife. Even ponds built for livestock watering are often designed for the other uses as well.

Livestock Considerations

Fence ponds to exclude livestock. The fenced area should provide plenty of grassy cover for wildlife nesting habitat and should include a permanent vegetation strip at least one hundred feet wide around the entire water line and at least ten feet from the downstream side of the toe or bottom of the dam. Don't allow trees to grow in or on the pond dam, but they should be encouraged to grow around the upper drainage portion of the pond.

Provide a livestock watering source below the pond. One of the most effective ways to do this, and one we use on our property, is to build a frost-proof concrete watering tank below the pond. This device provides animals with continuous water from the pond without having to pump it, and the water will not freeze in moderate weather. Ours sometimes freezes in single-digit temperatures, but it's easy to break the thin skim of ice each day.

Size and Design Considerations

Pond size and depth. A cold-water pond (with temperatures seldom rising above seventy degrees F. in the summer) for put-and-take rainbow trout (catch and keep, but no spawning) can be fairly small, about one-quarter to one-half acre. A pond stocked with bass and bluegill should cover at least an acre. Ponds for raising channel catfish or minnows can be smaller, even down to a quarter

Size is important when creating a fishing pond; it should be at least one acre in size unless the fish are to be fed artificially.

acre if the fish are constantly fed fish food. The water in a portion of the pond should be at least ten feet deep to prevent winter kill. It's not usually necessary to create ponds over fifteen feet deep because fish in warm-water ponds (in which temperatures range from eighty-five degrees in the summer down to thirty-eight or thirty-five degrees in winter) seldom use those depths because of the lack of oxygen.

A fishing pond designed primarily for bass should be constructed with little or no shallow water. Shallow water encourages weed growth and protects little bait fish from the bass. A minimum depth of two feet is suggested, with a fairly steep edge. Most ponds are constructed as smooth basins that provide little cover and structure for fish such as bass. If you leave creek and drainage channels intact and create structure in the form of underwater humps—and even islands and points around the pond—you'll provide more diversity and fish habitat as well as more general wildlife habitat. Geese and ducks will readily use islands that are at least twenty feet in diameter and that rise at least two feet above the water line.

The modern approach to creating a really good bass pond or lake is to create lots of underwater structures. Rock and boulder piles, standing timber and brush in the pond basin, even old car bodies and discarded

Bass-pond builders today incorporate many items to create fish-holding structure in ponds and lakes. These may be rock or boulder piles, old concrete drain tiles, underwater humps and islands, even a barrel filled with concrete to hold hardwood limbs. COURTESY RAY SCOTT'S VIDEO, CREATING GREAT SMALL WATERS

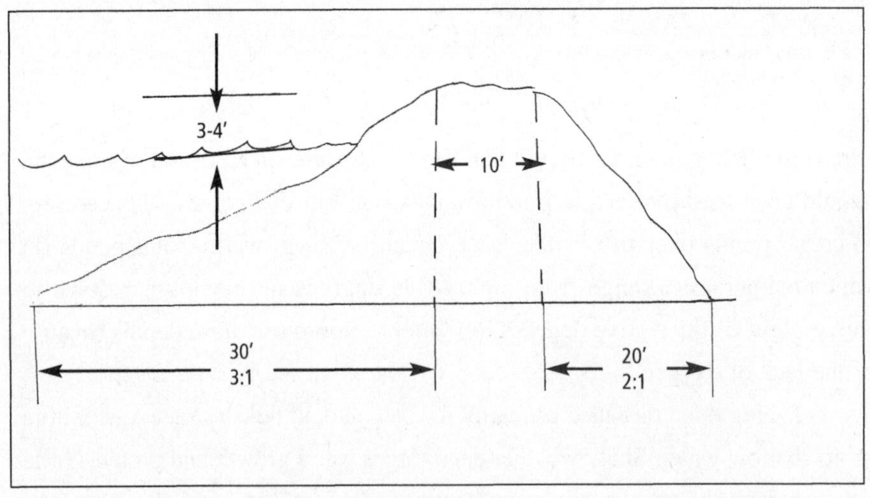

The design and size of the dam is extremely important in order to create a safe and productive pond with easy maintenance.

concrete culverts are used by serious bass-pond builders. Spawning sites can be created by building circles of concrete blocks and adding a two- to four-inch layer of pea gravel. These sites should be in water two to six feet deep depending on water clarity, shallow in murky ponds, deeper in clear ponds.

Dam and spillway size. The design and size of the dam and spillway depend on the type of pond construction, the drainage, and the area to be impounded. The top of a dam less than twenty feet high should be at least ten feet wide to provide a roadway for crossing and to prevent serious muskrat damage. For each five feet of dam height over twenty feet, the width should be increased by two feet. The slope of the front of the dam—the water side—of the dam should be from 3:1. The slope of the back of the dam—the land side—can be from 2:1 to 3:1. The first, 2:1 is the most economical to construct, but the 3:1 provides for easier mowing and maintenance.

Spillway design. There are several types of spillways: a principal spillway, a trickle tube, an emergency spillway, or a combination of either a trickle tube or a principal spillway and an emergency spillway.

Spillway design is also important. Several different methods may be used. A principal spillway and emergency spillway is one method. The ground spillway must be flat and allow for only shallow water overflow to prevent the loss of fish.

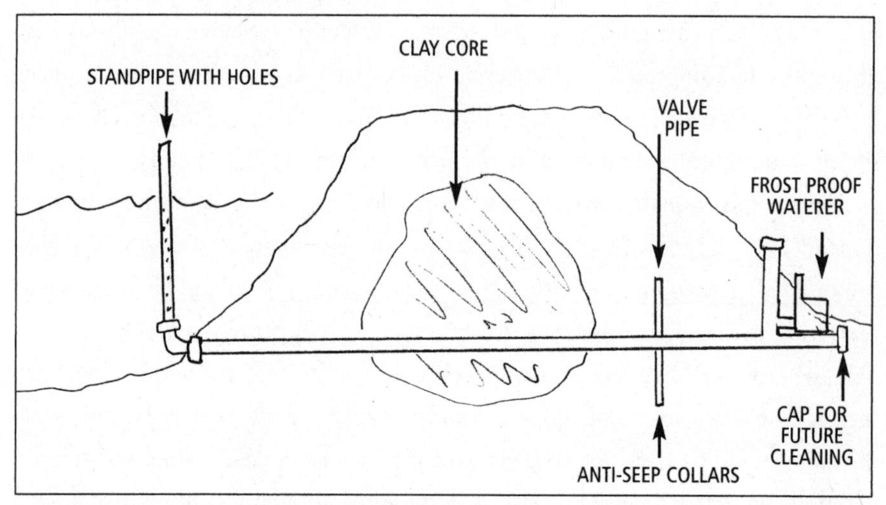

In most instances you should install a drainpipe in the pond. This allows you to control vegetation as well as fish populations.

- *Principal spillway.* It should be wide enough to allow floodwater to flow but shallow enough that big fish can't swim out. A common mistake made by pond owners is screening the spillway flow to keep fish from escaping. The screen traps trash, eventually collapses, and ruins the usefulness of the spillway. So don't screen—you'll lose lots of little fish, but they'll be replaced by natural reproduction.
- *Trickle tube.* A drainpipe that goes through the pond dam at the overflow and allows for excess water to run through without eroding the emergency spillway.
- *Combined spillway.* Make sure it has the capacity to carry stormwater runoff.

Draining the pond. You may wish to install a drainpipe for draining the pond, and in some states they are required by law. The drainpipe should be large enough to drain the entire pond within two weeks.

Stocking Fishing Ponds

In addition to creating a well-designed, successful pond, proper stocking is important to provide for fishing fun, raising fish for the table, or raising fish to sell. As always, decide on your goals and then create a management plan to accomplish them. The pond must be stocked with the correct numbers and types of fish and/or forage fish in order to be successful.

Wild fish. Don't stock with wild fish even if a pond already contains them.

A fishing pond designed primarily for bass should not have shallow-water edges that promote excessive weed growth.

If wild fish are already there, use rotenone to kill them before you stock with new fish. You will probably need a permit to use rotenone.

Trout fingerlings. These are stocked on a put-and-take (ponds in which the fish don't reproduce—they just grow) basis in cold-water ponds.

Bass, bluegill, and channel catfish. This is one of the most popular warm-water pond stocking combinations. The bluegills provide the forage base for the bass as well as fishing fun and food for the table. You may wish to add channel catfish for further fun and food. For waters of moderate to good fertility, the suggested initial stocking rate is a combination of one hundred bass, five hundred bluegills, and one hundred channel catfish fingerlings per surface acre of water. If you have lower water fertility, use lower stocking rates. The bass/bluegill combination will provide excellent bass and bluegill fishing for many years. In most instances channel catfish do not reproduce in ponds and must be restocked regularly, many times once a year. To accelerate the growth of the bass, add two to four pounds of fathead minnows per acre one growing season before the initial bass stocking. If bluegills have taken over an existing pond, stocking fifty bass in the eight- to twelve-inch range per acre can help, although it's better to drain the pond and start over. Catch and release all bass until the bluegills are under control.

What not to stock. Don't stock bullheads, carp, green sunfish, and crappies. Although the latter are favored by fishermen, they quickly overpopulate a pond with one- or two-ounce fish, reduce the bass population, and upset the bluegill/bass balance.

The most popular stocking numbers for ponds is one hundred bass, five hundred bluegill and one hundred catfish per acre. Bluegill provides food for bass and the table. Catfish must be restocked regularly.

Feeding Fish

Fish in most ponds don't need to be fed, and overfeeding can deplete the oxygen in the water and kill fish. In many cases ponds don't need to be fertilized either, although in soils with low fertility, fertilizing can increase fish production. A number of specialty pond fertilizers are available. Scotts 10-0-50 Timed Release Fertilizer is a patented once-a-year timed-release granular fertilizer for use in most sport-fish ponds. The suggested rate is twenty-five to thirty-five pounds per acre. Apply it in eighteen to twenty-four inches of water in the spring after the water temperature reaches sixty-five degrees F. One method of using granular fertilizer is to place it on wooden pallets in the shallow areas of the pond. The fertilizer bags should have holes in them and be submerged. Osmosis, wind and wave action will gradually disperse the fertilizer. Make sure the fertilizer is placed on a fertilizer platform or some other device to keep it off the pond floor. Monitor the pond with a secchi disk, a circular disk painted in black and white quadrants, or you can use a white coffee cup. Lower it into the water until you can't see it. This depth indicates the turbidity which translates into fertility. This works unless the pond is muddy to determine whether additional applications are needed during the year. The clearer the water, the less fertile. The best-producing ponds have water clear enough that a white object or secchi disk can be seen at least eighteen inches deep but not more than thirty inches deep in direct sunlight. If the color of the water is green, which limits sunlight penetration, you may need to reduce the amount of pond nutrients.

Keeping Records

Keep records of the fish caught in your pond and limit the harvest. Don't harvest bass for the first three years after stocking except for catch and release. The fourth year, harvest thirty bass per acre that are in the eight- to twelve-inch range and release twelve- to fifteen-inch bass. Bluegill and catfish may be harvested as desired, but the catfish may have to be restocked periodically.

Raising Catfish

Many landowners with ponds raise catfish for food or profit. The stocking rate is generally three hundred per acre in an unfed pond if no other fish are present. For maximum production, a quality catfish feed should be supplied regularly. If this is done, as many as two thousand catfish may be stocked per acre, with one thousand to one thousand five hundred a more realistic figure for beginning growers. Construct catfish ponds with draw-down facilities or sloped sides for easy seining. Some growers raise catfish in mesh cages.

Raising catfish for fun and profit is also popular. In this case the stocking rates are higher. For fed ponds, the rates can run as high as two thousand fingerlings per acre.

Pond Management

One of the biggest pond maintenance problems is vegetation control. Dense vegetation makes fishing from the bank difficult and swimming unpleasant. All vegetation is not bad—a certain amount of it is needed for good fish growth in a pond. Aquatic plants provide food for insects, which, in turn, provide food for fish. Plants also produce oxygen, provide cover for small fish, and protect shorelines from wave erosion. But a pond with too much vegetation can create more than just problems for fishermen. Periodic die-offs of dense vegetation can threaten fish. When a die-off occurs, it's usually after a period of cloudy weather. When large amounts of vegetation die and begin to decompose, oxygen is removed from the water and fish can be stressed and even die.

Controlling Aquatic Vegetation

Identifying the vegetation. To control aquatic vegetation, you must first identify it. There are four types: *Algae* is a small plant that doesn't have true leaves or flowers. *Floating plants* have leaves that float on the surface and roots that hang down in the water but are not connected to the bottom. *Submerged plants* grow underwater; they are rooted in the bottom, have stems and leaves, and produce seeds. *Emergent* or *marginal plants* are rooted in the pond bottom and have parts extending above the water's surface. Shoreline plants are included in this group.

Controlling aquatic plants. Aquatic plants are common in clear, highly fertile ponds with extensive shallow areas. If they occupy more than ten percent of your pond, they can be controlled in one of four ways: preventive, mechanical, chemical, and biological.

- *Prevention* is the best control method. Construct fishing ponds with shore areas at least three feet deep. Existing ponds with shallower areas can be dug deeper during periods of low water. High fertility may result from livestock runoff, agricultural fertilization or other sources.
- *Mechanical* control is effective on shoreline vegetation. Hand-pull cattails, willow trees, and cottonwood trees. Remove submerged vegetation by raking or pulling a chain through the pond between two tractors.
- Most *chemicals* affect only certain plants, so it is important to

identify the problem plant before purchasing a chemical. County agricultural agents and district fisheries biologists can recommend which chemical to use. Follow the label instructions explicitly. Most aquatic herbicides will not harm fish if used correctly. They are most effective when applied in April or May as vegetation begins to grow. If they're used later, treat only a portion of a pond at one time to prevent a massive die-off of vegetation that may result in a fish kill.

- One *biological* control method is the use of herbaceous fish such as the white amur or grass carp. A native of China, the grass carp does not reproduce in ponds. As it matures, its diet consists almost entirely of aquatic plants.

Water turbidity, aquatic vegetation, leaky dams, siltation, and other problems can make farm pond management a challenge. With a little research, however, and some technical assistance from state fish and game departments, these problems.

One of the most frequent problems with ponds is vegetation control. With management, the proper amount of aquatic plants will provide food for insects, produce oxygen, and offer cover for small fish. COURTESY FIDUCCIA ENTERPRISES.

CHAPTER 33

PREDATORS

Predators and pests are common problems with land managed for wildlife; you can't have one without the other. Control takes many forms. If you lease land from a farmer or other landowner, or if you own land, you can have a lot of fun and help control predators, varmints, and pests at the same time. Our waterfowl club lease is owned by a dairy farmer, and after the fall/winter waterfowl season is over, we have a club shoot that benefits the landowner. The shoot begins with a snow goose hunt. By March thousands of snow geese descend on his wheat fields, and he's more than happy to have us do wheat field guard-duty. Normally there's only one chance at the geese, which means that gunning for the main flock lasts about five minutes, although occasionally smaller flocks filter in. After two or three days of gunning, the geese leave the area. As soon as the decoys are picked up, however, the second shift starts—we simply head to the field edges and set up for crow hunting.

If you manage for an abundance of wildlife, you're going to have predators. With good habitat management, predators are not normally a problem. It does pay, however, to keep predators in check by hunting or trapping them.

Hunting Coyotes and Bobcats

The biggest problems for most wildlife, however, are predators such as coyotes, fox, and bobcats. They can play havoc with turkeys and deer. In fact, it's quite common for coyotes to respond to a turkey call. One season, in desperation to attract a huge gobbler, I was throwing every call in the book at him and he was responding in kind. During the fracas a female coyote came in and sniffed my boot! Unfortunately coyote season wasn't open in Missouri.

I did, however, see a coyote get his comeuppance from a big gobbler. Early in March I was filming the gobbler and several hens when a big coyote came in and sprang at the gobbler. The old bird rolled backward on the ground and hooked the coyote under the chin with his spurs, flipping the dog over. Then the gobbler started to strut again, while the coyote slunk off to lick his wounds. Unfortunately, I scared off the film stars with my howls of laughter.

A couple of years earlier, in February, I photographed a gobbler with the center section of his tail feathers gone. Apparently a coyote or bobcat got a mouthful of feathers instead of a meal. Both of these predators also do damage to young critters.

Many hunters and landowners in rural areas carry a coyote rifle in their vehicle in case they have a chance to shoot at a coyote. Typically, late winter sees the most coyote action, and some hunters spend Saturdays driving the farm roads and fields glassing for them. A flat-shooting rifle such as a .22-250 or even a .22 magnum is best for this type of hunting. Even though many landowners welcome coyote shooters, it's best to ask for permission before you hunt on someone else's land. Most of your shots will be in open, prairie-type country, and it's important to make sure you know what's in the line of fire behind the coyote.

Calling Coyotes and Bobcats

The most common method of hunting both coyotes and bobcats is calling. Predator calling not only helps to control the predators, but it's also, in itself, a great hunting experience. Here are some tips for hunting these animals:

Choose your call. Both can be called with mouth calls or electronic calls. Before you hunt, make sure you understand your area's seasons, limits, and legal methods of hunting. For instance, some states don't allow electronic calls or red lights at night.

Choose the proper calling location. The location should, of course, be in an area that has predators, as you can determine by scouting for sign, seeing the critters, or talking to landowners. You can also check the area by using a coyote howler to see if coyotes howl back.

Choose a high spot. Ideally, the location should be higher than surrounding land and have a wide-open field of view so you can spot animals coming in fairly easily. A calling spot located in dense brush can be utterly frustrating.

Choose a downwind spot. Your calling spot should be downwind of the area where animals are located or downwind from the direction they'll be coming from. Cagey critters will circle to come in downwind of the call, so try to find a spot with an open area downwind so if they try to circle they'll be out in the open.

Choose a site with cover for you. The area should have a natural means of concealment such as a brush pile, fallen log, treetop, or rock pile. Wear appropriate camouflage clothing and gear. If possible, put the sun to your back. This will allow you to see the coyote clearly, but it won't be able to see you as well.

Choose your approach. Make sure you approach the area as quietly as possible from downwind. If predators become aware of your presence before you start calling, most won't respond.

Choose your time. One of the most effective times to call predators is at night—with lights where they're allowed. The next best times are the hour before dusk and the hour after dawn. If possible, avoid days with high winds, as animals can't hear well and are easily spooked.

Consider using decoys. Coyotes and bobcats sometimes come to a call at a dead run, especially in areas where they're not hunted aggressively. At other times they'll stop just out of range and look for what's making the sound. A decoy can add realism to your setup and make your shooting more productive. The Feather Flex rabbit from Outland Sports is a soft foam decoy mounted on a motion-making machine. If you're calling with rabbit or hare sounds, the Feather Flex is irresistible to coyotes and bobcats. Some hunters add to the deception with a coyote decoy—either a homemade cut-out version or a foam target/decoy.

Use variety to remedy call-shyness. Using rabbit and hare distress calls has led to successful predator hunts for many hunters. But coyotes can become call-shy quite rapidly—they've heard too many of the same calls and have discovered that humans are the source of the high-pitched wails. Don't be afraid to try other calls—coyote howls and barks, pup squeals, woodpecker distress cries, and fawn bleats. All of these are available on high-quality cassettes for electronic calling.

Try combining calls. Alternate coyote howls or barks and prey distress calls. Begin by blowing a series of three to four soft howls with twenty-second intervals between. Allow three to four minutes to pass and then blow a series of soft distress calls each at least forty-five seconds long. Wait two to three minutes and then begin another series of howls, this time louder. Continue alternating calls for at least fifteen minutes, but no longer than a half hour. By alternating distress and howl calls, you may call in a hungry coyote or one who's curious about another coyote in its area. Starting with the soft calls won't alarm a closer coyote. If you happen to shoot or spook a coyote, don't stop the calling sequence. Other coyotes may continue toward the calls.

TRAPPING FURBEARERS

If your land has turkey, quail, pheasants or nesting ducks, their worst enemies are nest-raiding, egg-sucking skunks, opossums, and raccoons.

Furbearers, such as skunks, opossums and raccoons can be a real problem. With fewer trappers, furbearer numbers have jumped and they're death on the eggs and young of ground-nesting birds including many waterfowl. Furbearer trapping is a must part of today's wildlife management.

Unfortunately, with the price of fur at an all-time low plus the hard work that goes into trapping furbearers, less trapping is being done than in the past and many species are on the rise. Trapping furbearers is a wildlife management practice that can improve the numbers of game birds on the land you hunt. It's also an enjoyable way to make money from your land. Be sure to follow your state's or area's trapping regulations.

Controlling Farm and Building Pests

Starlings, house sparrows, rats, and groundhogs can be major pests to farmers and landowners, and .22-caliber rimfire rifles work quite nicely on all four. Use .22 shotshells inside buildings. A safer method to use around buildings and livestock is a pellet rifle. Both pump and gas cartridge models are available, and Crosman has excellent choices of both types.

Hunting Crows

Crow hunting offers a lot of sport in addition to helping to control these pests—and the shooting practice you get can improve your success with upland birds and waterfowl. Using calls—mouth or electronic—to attract crows is the most common hunting tactic. Electronic calls, such as the Lohman model from Outland Sports, provide the most realistic sounds for less experienced hunters. Also available from Outland Sports are their Fighting Crows over Baby Crows Distressed and Owl Hooting and Crow Fight tapes. A complete collection of Dennis Kirk's Crow Calling Sounds is available on cassette and include Crows Fighting, Crows Distress, Crow and Hawk Fight, Crow and Owl Fight, Crows Feeding (especially effective during the early morning hours), and the irresistible Death Cry of a Crow. Adding a Feather Flex owl decoy and crow decoys can increase your productivity.

With a varmint, predator, and pest control program, you can have more game to hunt and a lot of fun hunting.

CHAPTER 34

POACHERS

If you own land, you're going to have trespassing and poaching problems. I know! I've had fences cut, gates taken down, pastures cut up with tires, deer and turkeys poached, and items stolen. Chances are you've had some of the same problems.

Unfortunately, the more attractive your property is to fish and wildlife, the more problems you'll have with poachers. You can, however, make your property less inviting to poachers by keeping food plots, supplemental feeders, salt and mineral licks, and hunting stands away from property boundaries and out of sight of roadways. Keeping animals concentrated in areas away from roadways and boundaries will help discourage poaching and road-hunting. Likewise,

If you do everything right, you'll have lots of wildlife. But, you'll probably have lots of two-legged visitors as well. Trespassing and poaching are very serious and common problems, as shown by this poached gobbler with its breast and beard removed.

ponds and lakes that are not visible from the road are less likely to attract uninvited fishing.

Controlling trespassing—and the poaching that usually results—begins with an understanding of the trespass laws of your state. Fortunately, many states are passing stricter trespass laws and enforcing existing laws. Talk to someone at your sheriff's, constable's, or game warden's office about the laws and regulations that apply to you and your rights regarding trespass control. These officials should also be able to give you advice about the best methods of handling trespass in your immediate area.

Posting Your Property

The first tip they'll probably give you is to mark your property boundaries, but this can be a problem if you have unfenced land. A common sign I see in the Ozarks and other hill-country locations is a KEEP OUT sign, often sloppily hand-painted on an old tire nailed to a tree. *Deliverance*-type movies have made signs like this effective, although they can be an eyesore to property. In some states property boundaries in unfenced wooded areas can be marked with paint slashes on trees, or you may prefer to post NO TRESPASSING or NO

Understand the trespass and poaching laws in your state. Follow the laws, post your property accordingly, enforce the laws and prosecute.

HUNTING signs. Unfortunately, people often consider these signs as a dare to trespass. A more friendly, and often more effective, sign says FISHING AND HUNTING BY PERMISSION ONLY and has instructions for contacting the landowner to obtain permission. A friend of mine manages several thousand acres of leased deer land and says that since going to these types of signs his trespass and poaching problems have lessened.

A cousin solved the problem quite easily about twenty years ago. A string of "city hunters" continued to trespass on his land to hunt quail even after he tried everything from being nice and asking them to leave to posting his land with KEEP OUT signs. He even used his tractor to tow a car blocking his gate so he could get into a field! Unfortunately, the car belonged to a hotshot lawyer who caused him a bit of frustration. Then one day I happened by his place and saw new signs. They read HUNTERS WELCOME! HUNTING FEE $200 A DAY—DON'T WORRY ABOUT COMING BY THE HOUSE, WE'LL BE BY TO COLLECT. Twenty years ago two hundred bucks was a lot of money for a day's hunt, and paying to hunt quail in that part of the country wasn't even an idea yet. "Only collected once," Alvin related years later with a grin. "Sure helped with the fence repair bill, though."

Watch Groups and Hotlines

In many locations, the single best thing you can do to control trespassing is to join or set up a neighborhood watch. Watches are common in suburban areas, but they can be just as effective in the country. Get together with your neighbors and discuss how you can work together to solve trespass problems.

Most states have hotlines you can call to report poaching. Post the numbers near your phone and don't hesitate to use them.

Sometimes trespassing begins as a small problem, but it usually escalates into bigger problems such as poaching, theft, and property damage. Controlling trespassing is an ongoing problem that is never easy, but the above tactics can help.

Sources

API, Outland Sports, 888-530-6098, *www.apioutdoors.com*

Adams-Briscoe Seed Co., 770-775-7826, *www.abseed.com*

All Seasons Feeders, Inc., 800-841-1720, *www.allseasonsfeeders.com*

AMCO Mfg., 877-647-2563, *www.amcomfg.com*

American Hunter Feeders and Blinds, 972-285-7650, *www.gsmoutdoors.com/american-hunter*

Antler King Trophy Products Inc., 888-268-5371, *www.antlerking.com*

Acres of Antlers 'n' Acorns LLC, 502-753-6600

Audio Link Game Feeder, 800-330-0017

Bass Pro Shops, 800-BASS-PRO, *www.basspro.com*

Bombardier, 450-532-2211, *www.brp.com*

Brier Ridge Wildlife Products, 800-356-7333, *www.liftseed.com/brier-ridge*

Buck Busters Seed Co., 800-562-4570, *www.buckbustersseedcompany.com*

Buck Forage Products, 800-562-4570, *www.buckforage.com*

CDS, Inc., 800-791-1333, *www.cdsinc.ca*

Cabela's, 800-237-4444, *www.cabelas.com*

Craftsman/Sears, 800-377-7414, *www.sears.com/craftsman*

Crosman Corporation, 800-7-AIRGUN, *www.crosman.com*

Cycle Country, 800-841-2222, *www.cyclecountry.com*

Evolved Habitats, Pro-Graze Perennial Forage, 225-638-4016

Game Country, Inc., 912-883-4706, *www.gamecountry.biz*

Great Plains, 800-255-0132, *www.greatplainsmfg.com*

Hunter's Edge, 888-455-0970, *www.hardhunter.com*

Hunter's Specialties, 800-728-0321, *www.hunterspec.com*

Jonsered Tilton Equipment Company, 603-964-9450, *www.jonsered.com*

Kawasaki, 800-661-RIDE, *www.kawasaki.com*

Kenco, Outland Sports, 800-922-9034

Knight & Hale, 800-422-3474, *www.knightandhale.com*

Lohman Game Calls, Outland Sports, 888-530-6098

Long Agribusiness, WOWCO Equipment Co., 281-383-3100, *www.wowco.com*

Maxxis ATV Tires, 800-4-MAXXIS, *www.maxxis.com*

Megabucks, F.M. Brown's Sons, Inc., 800-345-3344, *www.fmbrown.com*

Monroe-Tufline Mfg., Inc., 800-537-5943, *www.monroetufline.com*

Moroney's Cycle, 845-564-5400, *www.jimmoroneyscycle.com*

Mossy Oak Bio-Logic, 888-MOSSY-OAK, *www.mossyoak.com, www.plantbiologic.com*

Moultrie Feeders, 800-653-3334, **www.moultriefeeders.com**

NovaJack, 800-567-7318, *www.novakjack.com*

Payeur, Inc., 819-821-2015, *www.payeur.com/en*

Pennington, 800-285-SEED, *www.penningtonseed.com*

PLOTMASTER, Woods-N-Water Inc., 888-440-9108, *www.theplotmaster.com*

PlotSpike Seed, Regan & Massey, 800-264-5281, *www.plotspike.com*

Purina Mills, 800-227-8941 *www.purinamills.com*

Remington QuikShoots, Superior Quality Products, Inc., 920-496-1039, *www.remington.com*

San Angelo/All-Luminum Products, 800-531-7230

Schuster Farms, 800-332-4045, *www.schusterfarms.info*

The Scotts Company, 800-811-2545, *www.scotts.com*

Stihl, 800-GO-STIFL, *www.stihlusa.cmo*

Sun Maxx, The Chaslyn Co., 417-866-2213

Sweeney Enterprises, Inc., 800-443-4244, *www.SweeneyFeeders.com*

Swisher, 660-747-8183, *www.swisherinc.com*

Tecomate Seed Co., 888-MAX-GAME, *www.tecomate.com*

TimberKing, 800-942-4406, *www.timberking.com*

TrailMaster, Goodson & Associates, Inc., 913-345-8555, *www.trailmaster.com*

TrailTimer Co., 651-738-0925, *www.trailtimer.com*

TrophX, Trophy Xcellerator, 888-675-4567

Whitetail Institute, 800-688-3030, *www.deernutrition.com*

Whitetail Strategies, 800-652-7527, *www.whitetailstrategies.net*

Wildlife Nutritional Systems of Texas, 888-355-WNST, *www.wnst.com*

Winchester Nutrition, 314-645-6494

Winchester Wildlife Feeders, B.A. Products, 800-847-8269, *www.feeders.com*

ORGANIZATIONS

American Forest Foundation, 202-463-2462, *www.forestfoundation.org*

Delta Waterfowl, 888-987-3695, *www.deltawaterfowl.org*

Ducks Unlimited Inc., 901-758-3927, *www.ducks.org*

National Wild Turkey Federation, 800-THE-NWTF, *www.nwtf.org*

Pheasants Forever, 877-773-2070, *www.pheasantsforever.org*

Prairie-Chicken Newsletter, Missouri Department of Conservation, or

your local Wildlife Management Biologist

Quail Unlimited, 803-637,5731, *www.qu.org*

Quality Deer Management Association, 800-209-337, *www.qdma.com*

Ruffed Grouse Society, 412-262-4044, *www.ruffedgrousesociety.org*

WHERE TO GET HELP

Conservation Farm Option

Local NRCS office (Natural Resources Conservation Services), *www.nrcs.gov*

Conservation Reserve Program (CRP), local Farm Services Agency (FSA)

Emergency Conservation Program (ECP), local NRCS office

Emergency Watershed Program (EWP), local NRCS office

Environmental Quality Incentives Program (EQIP), local NRCS office

Forested Flyways, 202-463-2462

Forestry Incentive Program (FIP), local Farm Services Agency (FSA)

Partners for Wildlife (PFW), U.S. Fish & Wildlife Service

Shared Streams, 202-463-2462

United States Department of Agriculture, 202-208-5634, *www.usda.gov*

Wetlands Reserve Program, local NRCS office

Wildlife Habitat Incentive Program (WHIP), local NRCS office